I started using a small laser pointer on my hip and gluteal region where I have a chronic muscle spasm. The laser pointer reduced the spasm and the pain significantly. Because the laser pointer brings down muscle spasms it also seems to reduce the pain that radiates down my leg. There are times when the pain is so bad that I can't sleep but then I use the laser pointer and it brings the pain level down enough that I'm able to sleep. It really has changed my life.

—Kristina Muhl, R.N.

How absolutely great this is! I suffered from a crippling inflammation of the tissue in my right hand, wrist, and forearm which made it difficult to perform certain aspects of my job. Over the course of several months I was able to see appreciable relief in my symptoms. I am now back to 100% functionality and able to perform my duties with no issues. If I feel occasional twinges of pain, I use a laser pointer at home for treatments on myself which are very effective. I also suffer from diabetic neuropathy in both feet. I have experienced a significant improvement in sensation, and I feel laser therapy has helped delay its progression.

—Mary Frances Richard, R.N.

After several days of self treatment my pain not only subsided, but I haven't had the same severity since. I now use my laser pointer whenever I feel pain in my body at almost any site. At 62 years of age I find I now feel better than when I did when I was working. I am able to keep any pain and discomfort under control with my laser pointer.

—Mary Causey, Pittsburgh Police Detective, (retired)
Former Cardiopulmonary Technologist

I was skeptical. However, I read Rejuvenation and applied Dr. DelRe's techniques on my sore rotator cuff. To my surprise I felt better and the results lasted. Dr. DelRe presents techniques that work with clarity and confidence. Rejuvenation illustrates simple techniques using light therapy and how to apply them with detailed elegance.

— Allen D'Angelo M.S.

My first laser back in 1986 was a 2.5 mW HeNe laser. It did not create any miracles in my dental clinic but the results were interesting enough to keep up my interest - in spite of the high price tag in those days. The years have passed and equipment for low level laser therapy has improved a lot, producing much higher power and new wavelengths. But still at a rather high price, especially if you want to have one for your private use. Meanwhile, laser pointers can be obtained at a very low cost; a red pointer of around 5 mW can be found for some 20-30 USD. Such sources of coherent light are by no means optimal but they do present an affordable option for self treatment of many medical conditions, especially for conditions in superficial tissues.

The aim of the book "Rejuvenation" by Lawrence DelRe is to underline this fact and to bring attention to the possibilities of self treatment by laser pointers. The suggested therapies are based upon the author's own long-time experience. Some indications are rather speculative, others based upon scientific evidence. Nearly 300 references are listed at the end of the book. But in all cases the author underlines safety precautions in a responsible way. The medical risks with laser pointer in this power range are indeed minimal but whatever there is has to be emphasized, considering the fact that this book is aimed at a general public of laymen. The author also underlines that laser pointers are for self treatment – no to treat others than yourself.

Although the range of possibilities with laser pointers is somewhat limited, many individuals will find them surprisingly effective for several conditions. This in turn may generate more interest in professional laser therapy and eventually advance the frontiers of this fascinating medical tool.

—Jan Tunér DDS

Interview of Joan Preston, (severe diabetic.)

Len: What is your problem?

Joan: I am a diabetic and I have thyroid problems. I have diabetic neuropathy, (numbness,) in my feet.

Len: How did the laser pointer work for you?

Joan: At first it didn't do anything. After a couple of weeks, I increased the dosage to 5 minutes per surface. In about a month the shooting pains in my toes went away and I could walk better. After about 6 weeks, I became distracted and quit using my laser pointer. Some time later the pains came back and I realized I hadn't been using the laser pointer, so I started up again.

Len: In your case, do you think it was the laser pointer, or something else in your life that gave you relief in the feet?

Joan: I think it was the laser pointer. I didn't change anything else.

Len: On a scale of 1-100, how much relief did you get?

Joan: About 80% relief. The toes still feel like there's numbness there, but the shooting pain has lessened.

—Joan Preston, Diabetic

REJUVENATION

Using the Power of Light to Increase Vitality, Energy and Healing

For over 40 Specific Remedies

Dr. Lawrence DelRe

Publisher Information:

Health-1st; 20 Bailey Avenue #3; Mt. Washington, PA 15211

1-800-340-4823

info@health-1st.com

First Printing 2009

Includes References and Index.

First Edition

ISBN: 9780615485362

Illustrations

†The information provided in this book should not be construed as personal medical advice or instruction as it is for educational purposes only. This book, the publisher and the author make no claims about prevention of, diagnosis for, treatments for, or cures of specific ailments or diseases. If you have a medical condition you should seek professional medical advice from your physician. This book is not a substitute for the medical advice, counsel and treatment of medical professionals. This book should be used only as reference tool to assist you. No action should be taken by anyone based solely on the contents of this book. While this book represents helpful information it is impossible for any book to be able to identify you or your loved one's medical problems or to make specific medical recommendations for you or a loved one. One's medical needs, specific treatment options, specific medical care, and specific treatment fees can only be determined after clinically seeking the recommendation of licensed medical professionals. Readers should consult appropriate health professionals on any matter relating to their health and well-being. The information and opinions provided here are believed to be accurate and sound, based on the best judgment available to the author, but readers who fail to consult appropriate health authorities assume the risk of any injuries. The author or publisher is not responsible for errors or omissions.

ACKNOWLEDGEMENTS

I would like to say that there was a lot of human interaction in my journey through this enlightening trek into coherent light therapy. Well, there was, regarding my patients. My wife, Eden, has been open to this, and I have learned much through her experiences. Also, there was Fred Kahn, M.D., and his great staff at his clinic in Toronto.

But that is it. Most of my interaction with professionals has been in the role of simply informing them about it.

Not only would I like to thank those, mentioned above, but I owe a great deal of thanks to Jan Tunér and Lars Hode, who co-authored "Laser Therapy". I consider their book to be a classic on coherent light therapy for the clinician.

I would like to thank Tiina Karu for her long dedication and contribution to scientific research into coherent light therapy.

I would like to thank my wife, Eden, for her invaluable contribution to this work. My son, Nathan, provided the wonderful illustrations in this book. Great job, Nate!

I would like to thank Bookcovers.com for their help in making this book as beautiful and professionally done as it is.

This book would not be possible without the hundreds, perhaps thousands of M.D.'s, PhD's, DDS's, DMD's, and other researchers from other disciplines, from around the globe, who have done the hard work and then had their research on coherent light published.

TABLE OF CONTENTS

TABLE OF FIGURES

INTRODUCTION

I. ABOUT THE AUTHOR

Dr. Lawrence DelRe has been treating the spinal and musculoskeletal complaints of his patients for eighteen years. He is a licensed Doctor of Chiropractic with degrees in Human Biology and Electronics. Dr. DelRe, who is fifty-three-years-old, has three children with his wife of twenty-eight years, Eden. A stateside Vietnam-Era veteran, Dr. DelRe is also a Mensan. In his short six years in the U.S. Air Force, he worked in the base dental clinic as the sole surgery tech and went on to run the radiology section at the age of twenty. After further training by the U.S. Air Force, he went to work in a precision measurement electronics calibration and repair laboratory facility in the Philippines.

While in chiropractic school, Dr. DelRe was skeptical of many of the unconventional therapies his fellow students were studying. Although the curriculum was formal, he, as well as most of his fellow students, also studied outside the school's curriculum. Some, for example, studied techniques such as iridology, the reading of the irises of the eyes to diagnose disease. Other techniques involved touching *special* spots of the body with certain fingers for the healing of disease. A few others believed that shining certain colors on the body would have healing properties. He viewed techniques such as these with skepticism. Over time, however, he saw that some unusual treatments could actually help patients at times. He learned how not to judge something automatically as nonsense.

In 1992, he patented a spinal segmental disc shear-resistance motion analyzer (Patent # 5,101,835). He developed and built the robotic version of it and has had his research with the robotic version published in conjunction with Ray Brodeur, PhD, DC, a researcher at Michigan State University Biomechanics Department.[1]

In his practice, Dr. DelRe received advertising for all kinds of devices and *"magic bullet"* nutritional supplements. He saw that when a new health idea came along, if it had merit, it would still be around and growing after five years or so—and so it was with coherent light therapy. In the beginning, he received the ads and threw them in the trash with the others, considering coherent lights as weak and unproven.

Eventually, Dr. DelRe happened to be at a seminar in which the speaker spent some time talking about the latest coherent light research. Dr. DelRe was impressed by the research and technology involved. One of the vendors offered a 100-milliwatt (1/10th of a watt), 830-nanometer (invisible red wavelength), portable, *cold* (low-level, unfocused), coherent light, which cost $6,000.00. Dr. DelRe did not want to spend that much on a device that was unproven in his experience. Their device was made in Europe. At that time, many therapeutic coherent lights came from Europe. Now they are made in many countries.

He did more looking and could not find a reasonably priced coherent light with the specifications noted above, so he designed and built a small, portable one for himself.

It was a 100-milliwatt, 830-nanometer coherent light and still works great. Since he had been a calibration tech in the Air Force, he made sure he had a good device to calibrate it.

Coherent light therapy is not new to the Food & Drug Administration (FDA). In fact, they created a completely new product code (NHN) for those coherent light devices in 2002, primarily based on a General Motors study on carpal tunnel syndrome. Since that study was conducted, it is said that General Motors made coherent light therapy available for most of their factories. Between 2002 and 2004, the FDA granted 510(k) approval to several companies to market lasers that provide low-level laser treatment (LLLT). The LLLT lasers are classified under "Lamp, Non-heating, for Adjunctive Use in Pain Therapy."

Dr. DelRe built two other coherent light units, made by combining 5-milliwatt visible red lasers (630 nanometers) and found different uses for them. He has used every common physical therapy device (ultrasound, micro-current, electric muscle stimulation, massagers, ice, heat packs, diathermy, etc.) out there for over eighteen years and has never seen anything work like the coherent light. He has been able to duplicate much of the positive research that has been published.

1   R. R. Brodeur and L. DelRe, "Stiffness of the thoracolumbar spine for subjects with and without low back pain," *Journal of the Neuromusculoskeletal System* 7(4) (1999):127–133.

Introduction

Dr. DelRe went to Toronto to visit a light therapy clinic that a doctor who had once been a surgeon owned and operated. About thirteen years prior to that meeting, the surgeon had suffered a bad rotator cuff injury that had not healed, and he was facing surgery. He did what he could to avoid surgery. The surgeon went to a Russian doctor there in Toronto who used red light on his shoulder, and it healed up. You will see later that the Russians are very advanced in this technology. They have been developing monochromatic light therapy for eighty years and monochromatic coherent light therapy for the past thirty years. They have a government-sponsored department dedicated to coherent light therapy.

The surgeon later started the light therapy clinic. He has a *teaching weekend* every month or so. Dr. DelRe saw before/after slides of cases of many conditions. Of particular interest were the cases of amputees who were about to lose their remaining leg, and they were saved in this clinic. The doctor there had a rather large staff from all over the world, and they were treating patients. Dr. DelRe was permitted to talk with the patients about their treatment and results. It is significant that Canadian national health insurance covered none of the treatments. Then he took a walk during lunch and was in front of the building when a man in the parking lot asked Dr. DelRe if he would help get his wheelchair out of the trunk. The man in the car was one of the case histories/slides that had previously been reviewed. This man had Buerger's disease and was one of the amputees in the cases Dr. DelRe had seen. The patient had recently dropped a soda can on the foot of his remaining leg, causing a new ulcer that needed treatment. The other amputee cases discussed were diabetic patients facing another amputation. He saw cases of rheumatoid arthritis, something he could not successfully treat prior to using the coherent light, successfully treated.

After using the coherent light in his clinic and at home, he has found, as many others have, that this is a method of bio-stimulation unlike anything else on the planet. Dr. DelRe believes that the world should not have to wait for this information. Coherent light therapy has been shown to be safe for the past thirty years. That is the reason for this book. He wants to bring coherent light therapy to you, now, not one hundred years from now, in your own home.

II. ABOUT THE BOOK

The Secret the Medical Industry Does not Want You to Know. That would make a great headline, wouldn't it? The only problem is that it is not true. The truth is that *they* do not know. If they did, they might try to squash it with de-bunking "facts." * But they just do not know. They *could* know, if they paid any attention to over 2,400 research articles that have been published for more than thirty years, which cumulatively demonstrate that illumination with coherent light therapy is harmless (*when used as directed*), yet can be effective[†] for

Muscle problems[†, 1]
Nerve problems[†, 2]
Tendon problems[†, 3]
Wound healing[†, 4]
Heart problems[†, 5]
Gall bladder problems[†, 6, 7]
Intestinal problems[†, 8]
Health of the cervix[†, 9]
Rheumatoid arthritis[†, 10]
Fibromyalgia[†, 11]
Breast tenderness[†, 12, 13]
Dental bone problems[†, 14]
TMJ syndrome[†, 15]
Carpal tunnel syndrome[†, 16]
Bursitis[†, 17]
Tendinitis[†, 18, 19]
Low sperm count[†, 20]
Synovitis[†, 21]
Prostate problems[†, 22]
Herpes[†, 23]
Stroke[†, 24, 25, 26]

Why do I use the term *coherent light*? Because I want to make a clear distinction that we are *not* talking about *laser beams*. Laser beams have the connotation of high-powered intense beams of focused light that can heat, burn, cut, and blind.

Coherent light has the same source, but I am using the terminology here in order to define it as unfocused, which means that it spreads out, like a flashlight, dispersing the energy in a softer, more *buffered* way. This form is less harmful to the eyes than a beam, although you must still be careful not to get it in your eyes. It treats a larger area and is very bio-stimulating. Therapeutic coherent lights use low power. As I will repeat throughout this book, *a 5-milliwatt laser pointer produces only 5/1000 of one watt.* At these low powers, the coherent light has virtually *no physical effect* on molecules, cells, and tissues; however, it has remarkable *physiological* effects on living cells and tissue processes. It *surprises* them. In all of evolution, your cells have never *seen* coherent light, and it triggers a healing cascade of reactions.

This book is grounded with the concept of *low-dose, low frequency of treatments with consistency.* The dosage/times I give are conservative and low. The treatment frequency is twice a week at the most, except for warts, which are every other day. I limit glandular treatment to once per week at the beginning and give numerous warnings of the known and unknown risks. This gives you (and your doctor) plenty of time to observe and adjust the dose, if necessary, or quit altogether if you have a negative result, such as temporarily aggravating osteoarthritis.

Focused lasers have a lens that can create a *laser beam* with pinpoint heat just as a magnifying glass does and are typical of the *laser beam* we all know about. In powers over 5 milliwatts, they become increasingly more hazardous to eyes. At higher powers, with a focused beam, it takes less time to do harm. At the low power of a 5-milliwatt laser pointer, which has a lens, the human blink reflex renders an inadvertent flash harmless.

Unfocused therapeutic coherent lights do not heat the skin or anything else in the body. There are no x-rays, ultraviolet light, or anything like that. They are somewhat safer than focused lasers regarding eye hazards because there is <u>no</u> lens-focused concentrated beam. They still have the characteristics of this very special light, though. We all know that diamonds are built with countless identical carbon atoms, which all *fit* together perfectly. I like analogies, so here is one:

CARBON IS TO A DIAMOND AS LIGHT IS TO A LASER BEAM

I guess you could say that coherent light is *hard* light. Coherent light has photons *fitting* together in lockstep. Laser therapists around the world use a wide array of different types of coherent lights of different wavelengths. They can also use pulses of various frequencies of those coherent lights for different conditions. The pulsing of coherent light *surprises* the cells even more. Think about a strobe light waking you up, compared to a light bulb simply being turned on. Pulsing the light may give better results but is not necessary in most cases. If you experience positive results with your self-treatment but are still *not quite there*, you should consider seeking a professional, such as a chiropractor or physical therapist, who uses therapeutic laser therapy. It would be nice to recommend a physiatrist also, but they do not seem interested. That is OK, for reasons I will go into later.

We will not be so sophisticated. We will be using visible red *constant* coherent light. It is a classic therapeutic wavelength. It has been repeatedly proven effective and safe for many years before the pulsed coherent lights came out and continues to succeed well. One form of it is commonly known as a standard red laser pointer.

It is my hope that you will use these therapies on any of your chronic complaints covered in this book. I hope that you will add years to your life and life to your years by using the rejuvenating information contained here.

The back cover of this book includes the term "Medical Freedom."

One of the signers of the American Declaration of Independence, Dr. Benjamin Rush, has this quote attributed to him: "Unless we put Medical Freedom into the Constitution, the time will come when medicine will organize and restrict the art of healing to one class of men, and deny equal privilege to others, this will be to constitute the Bastille of Medical Science. All such laws are un-American and despotic and have no place in a Republic. The Constitution of this Republic should make special privilege for Medical Freedom as well as Religious Freedom."

Medical science has served honorably to remove most superstition from the field of healing. Medical science annihilated Vitalism, which is a name given to the healing approaches that relied on a holy, spiritual, or natural force involved in healing. Vitalism was prey to superstition and charlatans. This does not mean that the concept does not contain truth.

I recently read a quote by a scientist. I was struck by what he said in a few short words: *"a scientist is a 'professional skeptic.'"*

I believe that is a wonderful, simple clarification.

If you were to ask your doctor if he considers himself a scientist, he would probably say yes. At the very least, he was a scientist in medical school. 75% of the chiropractic curriculum is the same as medical school. Sometimes people underestimate the underdog.

This book is based on science, but it is not scientific. It does make a leap. I believe I offer safe speculation. It may take one hundred years for well-respected scientists to accept the data as scientifically sound. I think that is too long for something that has been demonstrated to be safe.

There is so much knowledge about *getting better* that has been lost with time. There are many tools left out of your doctor's bag because of the confines of medical science. That does not necessarily mean the tools are bad or do not work. It just means he/she does not use them because those remedies are outside the medically proven models, which have adjustments each year.

People today have many more choices. More and more, it is up to you to decide about things such as nutrition, acupuncture, meditation, etc. It was not that long ago that a survey of 181 cardiologists revealed that 44% were routinely taking antioxidant vitamins. Among the supplement users, 90% were taking vitamin E, 75% were taking vitamin C, and less than half were taking beta-carotene. Although 44% of the cardiologists used antioxidants themselves, "a somewhat smaller percentage (37%) recommended antioxidants routinely to their patients."[27] *As scientists*, there was, and there still is, not enough proof to recommend vitamins. *As humans*, "why not, what could it hurt?"

Since starting this book, I may have run into possible *debunking* by people who may want to come up with research that shows coherent light therapy does not work. In order to use research to *find something wrong* with a competitor, you need to find a researcher that will give you the ammunition you need to limit that competition. Dr. (Name not shown) may have been a good man to use. Most, if not all, of his research into coherent light therapy ended up with a conclusion that it did not work, even on conditions that *are well known* to respond dramatically to coherent light therapy. These were a few of his sponsors: the Swedish Society of Medicine, Royal Swedish Academy of Science, and Swedish Association of Registered Physiotherapists.

The Dean of Research of the Karolinska Institute, Jan Carlstedt-Duke, wrote this explanation when (Name not shown) was investigated for scientific fraud and malpractice:

Following a number of allegations made 2002–2004, I carried out an investigation of *Dr (Name not shown)* and his group. The result of this investigation was a *clear indication of scientific fraud and malpractice.* We therefore asked the Swedish Research Council's expert group on scientific malpractice to make an independent investigation, which, due to the complexity of the case, was completed on May 31 this year. The conclusion of this investigation was that scientific malpractice was demonstrated with numerous examples of plagiarism within scientific publications and patent applications. Furthermore, the documentation of the research carried out and its description was so poor, and in many cases lacking, that it was impossible to determine if and when experiments were carried out or by whom. *The conclusion of the Rector is that we completely lack confidence in (Name not shown) as a scientist.*

(Name not shown) resigned from his position as professor at Karolinska Institute in October 2003. He has not been associated with Karolinska Institute since that time and has no affiliation with the university today. He resigned his position following confrontation concerning a manuscript that was used for a patent application and that was plagiarized from a published paper from another group and that

contained manipulated data that could not be verified with documented primary data. All connections with Karolinska Institute were severed following that incident.

There was an article in the Journal of the American Medical Association (JAMA), which gave numbers that add up to 225,000 Americans dying each year because of iatrogenic causes.[28] That means a medical doctor, nurse, drug, or hospital caused the death because of an error. It is the third leading cause of death in America. That is the equivalent of a 9/11 every ten days. I think you should know the numbers.

What am I trying to say? You should think for yourself if you can. You *can* get better with the low risk methods listed in this book. You can stay better if you want. What am I trying to say? You can study, take control, and treat many conditions on your own. You have the opportunity to learn and take your own risks. You cannot have freedom without risk. You can add your own common sense to trust, instead of always someone else's.

I think that is a form of medical freedom.

1

THE FUTURE

ONE HUNDRED YEARS

Please allow me give you a glimpse into the far future; let's say one hundred years from now. Endocrinologists will be using coherent light as one of their primary modes of treatment. They will stimulate the thyroid in low thyroid conditions. They will use coherent light on the adrenal glands and perhaps the ovaries to ease the transition of menopause.

They will eventually hand the work over to licensed coherent light therapists by prescription. Tonsillectomies will *always* be followed with at least one thirty-second (or less) dose of coherent light. Many sinus problems will be treated with coherent light before drugs are tried. Ulcerative colitis will have its coherent light protocols, starting with very low doses, followed by increasingly normal doses of coherent light. Rheumatoid arthritis will be nothing more than an inconvenience; most will not need the drugs we use for it today. The pancreas of diabetes mellitus patients will *always* be treated by coherent light because coherent light therapy will always decrease the need for insulin or drugs, if not eliminate the need completely. The male hormone testosterone deficiency will be boosted by exposing the testes to coherent light.[29] Local growth factor production will be increased for the elderly and for body builders by treating their muscles with polarized light or polarized coherent light.[30]

Many others and I honestly expect that many of the above predictions will be proven over time. Throughout the world, much of it is being done right now. The fact *is* that you can rejuvenate things that you have not thought about just because

For those with genetic markers for cancer, coherent light will routinely be used for *prevention* of:[†]

- Breast cancer [†]
- Cancer of the cervix [†]
- Prostate cancer [†]
- Intestinal cancer [†]

Here are some more conditions in which coherent light will prevent onset or worsening:

- Heart disease [†]
- Alzheimer's disease [†]
- Stroke from aneurism [†]
- Root canal avoidance and prevention [†]
- Alveolar regression [†]

In one hundred years, coherent light therapy will dominate:

- Geriatric medicine [†]
- Physical medicine [†]
- Endocrinology [†]

In one hundred years, coherent light therapy will be always used following:

- General surgery [†]
- Wound healing [†]
- Burn treatment [†]
- Plastic surgery [†]
- Dental implant surgery [†]

it did not seem possible. You can treat many of the "itises" you may be experiencing. You *can* become younger and stay younger longer. Healthier.

RISK

What are the risks? Although there are no known cases of eye damage using low-power therapeutic coherent light, you must not let coherent light into the open eye, just out of common sense. There are usually known and unknown risks with just about any treatment. I will cover the known risks at this time, as best I can.

Regarding the unknown risks, I believe that the past thirty years of the use and detailed research of this technology at much higher dosages than I recommend, without finding harm, is a valid indicator of low risk. (One paper indicates that the glands may become exhausted if they are directly illuminated 500% more often than I recommend in this book.[31] Please see Chapter 3: Safety.) There have been a number of specific studies published that *try* to do harm with coherent light,[32, 33, 34] and it just does not do even minimal harm unless used in dosages hundreds and thousands of times larger than the small dosages and conservative frequencies of treatment that I recommend in this book. In other words, coherent light therapy appears to be safer than any drug at this time when used as directed.

I rate each treatment protocol in regards to "Confidence" and "Risk." The important thing is to read and follow directions. Do not deviate from the low dosages I recommend in this book. Pay attention to any sensitivities or adverse effects. If you have been told you have mental problems/issues or if you are self-sabotaging, please do not read this book. Seek a professional who does coherent light therapy, usually a chiropractor or physical therapist. If you want to see a medical doctor who uses coherent light therapy, I would advise doing a computer search. If you can afford it, you can go to Europe and find more doctors with longer experience. I would look at Sweden first. There is also an excellent facility in Toronto, Canada, run by Fred Kahn, MD. He is a researcher and pioneer in the industry. The name and address is Meditech International Inc.; 415 Horner Ave., Unit #1; Toronto, Ontario, M8W 4W3.

Although I love the metaphor of *healing with light*, coherent light is not "the light that heals". It is not a substitute for spiritual light.

If you are extremely frail, please keep in mind that this technology is new, and what I talk about in this book may upset your system in ways we just do not know. See a specialist who uses coherent light therapy. If, however, you are reasonably healthy and have the ability to follow simple instructions and stay within guidelines, then please, you are welcome to join thousands of others on this journey into the future. Hop on board, and enjoy this ride. Your grandchildren will probably go to licensed coherent light technicians working under doctor supervision/prescription to get this treatment some day. One of them may become a licensed therapeutic laser technician him/herself because of the knowledge you have handed down to them.

2
A Closer Look at Coherent Light Therapy †

Osteoarthritis †

You may have noticed I left out osteoarthritis, the most common form of wear-and-tear arthritis, from the previous lists of chronic problems. In spite of the fact that some of the research shows a benefit of coherent light therapy for osteoarthritis,[35, 36] my experience has been that it does not help most osteoarthritis.[37, 38]

At this point, I believe that it causes such a significant bio-stimulating response that it is just too much for most joints, even in very small doses, and can cause a temporary acute inflammatory *flare-up*. Some authorities state that these joints can be treated by using very small doses until a person acclimates to the treatment. I have a patient who did have success with a thumb. You can try it if you like but start with low dosage times.

Rheumatoid Arthritis †

Rheumatoid arthritis, however, "loves the laser." It is simply incredible how quickly and effectively it helps that disease.[39, 40, 41] I have found that these patients can take normal doses from the start with no flare-up! I frankly do not understand why they respond so differently from osteoarthritis patients. They respond with more mobility, and pain decreases to a minimum if it remains at all. In addition, with the coherent light, they need less and less of it over time. It is possible to stretch treatment gaps to six weeks, maybe more, with time and dedication. One more thing, the coherent light appears to have a *systemic* effect on rheumatoid arthritis. This means that when you have been treating one or two joints with coherent light, *all*

5

the joints of the rheumatoid arthritic patient start *improving* as well. When we treat any areas in the body, we also treat the bloodstream at the same time.

TREATMENT OF BLOOD †

Eighty years of Russian and Eastern European research demonstrate repeatedly the amazing effects on blood that has been exposed to various monochromatic light wavelengths, in the early years, and to coherent light, in the later years. In his book, "Laser Irradiation of the Blood", Levon Gasparyan says, "Unlike the treatment mechanisms of local laser therapy, the medical effects of photo-hemotherapy methods are determined by predominance of systemic healing mechanisms above the local ones, increasing the functioning efficacy of vascular, respiratory, immune, other systems, and organism as a whole."[42] It will take years of research to prove these things conclusively if you want to wait. Considering the gaping lack of any evidence of harm, I do not see a pressing need to wait. A number of the treatment protocols in this book are directed towards *treating the blood flow leading to deeper tissues* we are trying to reach. You will be made well aware of when we are attempting such. †

The fact is you can cure yourself of many common problems, old injuries, and diseases if you follow the simple instructions in this book—all without piles of vitamins, herbs, drugs, exercises, surgeries, etc. (Although you do not need them to make these coherent light techniques work, you should still exercise and take some nutritional supplements/antioxidants/tea for general purposes.)

In this book, I will share with you what I, and many other professionals from all disciplines, have discovered over these years regarding coherent light therapy and chronic physical problems. The areas I will cover which are outside of chiropractic will be based on multiple researched studies and my personal experiences at home. A number of treatment protocols in this book are simply educated guesses, based on the research.

While the conditions related to the neuromusculoskeletal system have been treated with coherent light in my practice for years, other conditions that would be in the medical realm, such as glandular rejuvenation, have been used on me, and my wife has used the coherent light on herself for the same reasons as well.

I am interested in spreading the truth about this important information and increasing the use of coherent light therapies in our health care system. I do not want my fellow humans to have to wait one hundred years for this wonderful, drug-free, natural rejuvenating therapy to be available only to their great-grandchildren. I hope you are one of the ones who tries these rejuvenation techniques and that you tell your doctor about your treatment and your results, even if it appears he is not interested. Doctors do not usually accept new ideas from patients, but we need to help them to help others. If they really did "know it all," we would not have the millions of chronic pain patients with the same physical problems today that they have had for hundreds of years, would we?

Yes, we now have ibuprofen, which is wonderful, along with the other *profens*. We have come a long way in conservative medical treatment. On the other hand, *holistic wellness* and *rejuvenation* have always been in the realm of the adventurous, the medical heretics, the "nutrition nuts," and such.

Medical doctors tend to be conservative. I do not completely understand the culture, but new ideas are shunned, unless they come from the drug industry, or they come from a highly prestigious source or maybe a trusted medical colleague. It must come from the *top down*. It stems from working with certain patients that could die very easily. Who would want to take risks outside the medical models?

Regarding *life and death issues*, "the buck stops" with medical doctors, and I respect that. When patients want to do something outside of their doctor's experience, the doctor's first impulse may be to control it and redirect them into his/her scientifically proven paradigms. I am just analyzing the process.

If you are reasonably healthy, I am proposing to you now that you answer any *off-the-cuff* discouragement about coherent light therapy from your doctor. He/she is doing what he/she should be doing—being the guardian of the medical model.

Respond with specific questions based on this book. If you read this entire book, you will know much, much more than he knows about any of this, so do not be afraid to ask him for specifics, and you will be satisfied as to his/her knowledge, or lack thereof, about this technology.

If you want to check out if what I say is exaggerated or not, type "low level laser therapy research" in any search engine, such as Google. There will be so much information that it will keep you busy for weeks or months tracking it all down. There are many laser therapy books for professionals on the market. The book you are reading is meant for the general public as well.

"If you throw enough mud against a wall, eventually some of it starts to stick." My aim is to show, from the *bottom up*, that coherent light therapy is safe and works better for the conditions listed here than most of the remedies/treatments out there for most people.

The primary differences between the coherent light units I use in my office and what you can buy and use at home are in depth of penetration and time. It is all a matter of the math. You can buy the laser pointers almost anywhere for $3 to $25, and they will work for many problems that are within about ½ inch (<1 cm) deep from the surface of the skin. You can lightly press the coherent light into the skin and muscle for deeper penetration. *Do not press too hard.* I just had a patient today that told me that his relative used a pointed (not flat-tipped, as I insist) laser pointer on her arm and ended up with red marks because she pressed in too hard. Do I have to say it? Yup—do not do that. Buy a flat-tipped laser pointer.

The readily available red laser pointers have a lens, and thus, they are focused, however, the very tiny power level of 5 milliwatts (half of 1% of one watt) means that there will be no heat. In addition, you will get deeper penetration with a lens at this low wattage. To reach deeper structures, we can do this indirectly by treating the blood supply going to those structures. The blood carries the effects of the coherent light exposure.[43] The Russians have studied this phenomenon for eighty years and use it today. This would be my approach to Alzheimer's prevention[44] as well as other self-treatments.

In one study, red blood cells that were illuminated by coherent light were more flexible and stronger. Red blood cells are called *erythrocytes*. Erythrocyte deformability and erythrocyte ATP (energy) levels were significantly higher.[45] *Deformability* means that they will bend in and out of tight spots better. *Higher ATP levels* from illumination mean that the mitochondria in your cells actually act in a way *like chloroplasts—they have been shown to have receptors that convert coherent light into useful physical energy, as a leaf does.* This effect is one of the most repeatedly documented events of coherent light therapy. It is similar to recharging a bunch of microscopic batteries. Another analogy is *aerobic exercise for cells.* It is covered in Chapter 5: Physiology of the Healing Cascade. The white blood cells are affected in positive ways as well.[46, 47] When the blood stream is illuminated by coherent light in diabetics, the blood sugar level has been shown to decrease, along with the need for drugs. LDL cholesterol and blood lipid levels were also decreased significantly in the diabetic in another study.[48]

Your own coherent light pointer (laser pointer) sessions will take longer than my office coherent light will, but not too long. My basic rule of thumb is to hold the 5-milliwatt coherent light pointer against the skin and move it around within a 1-cm circular area for sixty seconds. That will give a mild dose (0.3 joules/cm^2) for that spot. For an area the size of a golf ball (3 cm^2), you would use three to five minutes of illumination. I believe that since the 5-milliwatt coherent light is held there much longer than my 100-milliwatt coherent light, the increased time of exposure for the 5-milliwatt light will result in less joules needed.

Less is always better, wouldn't you agree? Using just a little coherent light keeps the body's components all *in the loop*, instead of simply chemically manipulating them. This way, the body still has more control of its own homeostasis. Homeostasis describes a normal *balance* of all the systems of the body, working as one. For example, testosterone abuse (the *steroids* of athletes and bodybuilders) has been shown to cause shrinkage of the testicles. Why? Because the negative feedback loop tells the pituitary gland to stop making FSH, which is what makes the testicles produce testosterone. Apparently, without FSH, the testicles shrink. Although it is possible that it could happen because of the increased testosterone from illumination of the testicles, I believe it is less likely because the body is kept in the loop. Nothing has been added; the testicles are just rejuvenated. This is explained in more detail in Chapter 12: Male Health.

One other thing, this book is meant for people who have chronic problem areas that can use rejuvenation. It is not for people under twenty-one years of age, unless they have a specific disorder listed in this book or in another book about laser therapy. When you are young, your body is building, and you should avoid any *crutches* or enhancers. You will retard something.

Is there a limit to the number of spots treated in one session? There is no solid evidence on this. Some say that the first sessions should be limited to a total body dose of 20 joules. Using a 5-milliwatt laser pointer, this would equate to sixty-five different points treated for sixty seconds each. It is unlikely you would need to treat this many points or have the time to do so. By then, every blood cell in your body will have been illuminated, and that may be too much. The presence of a focusing lens is not important in this case because 5 milliwatts (5/1000 of a watt) is less than 1% of 1% of the power of a 60-watt incandescent bulb. It will not build any heat in the skin. Yes, I know I am repeating myself, but I want certain concepts to sink in.

For comparison, a 100-milliwatt, 650-nanometer coherent light would effectively penetrate 3–4 cm, (around 1½ inches.) You would use a treatment time of ten to twenty seconds per session over an area of 2–4 inches in diameter. That would give you 1–2 joules of energy spread out over that area, which is a mild dose.

Treatments for most chronic complaints are every three days, or twice a week, to start. Start at one treatment per week for glandular treatments. Specifics are given in the later chapters. In any case, the goal is to stretch it out to six weeks between treatments.

Allow me to give you an example of one aspect of what the coherent light does. An unhealed rotator cuff (shoulder) injury has sick, morbid cells that have tried to heal for weeks, months, or years. Those cells have *given up* trying to become normal and have accepted a *lower level of existence.* (Rather than give all scientific jargon, I will attempt to paraphrase and use analogies throughout this book.) Morbid cells have a number of physiological characteristics that demonstrate that they are unhealthy. One of the common characteristics morbid sick cells have is *increased cell membrane resistance.* This means that nutrients and such cannot get through the cell wall to wash out the garbage and supply fresh nutrients. Coherent light has been shown to drop cell membrane resistance instantly to a more normal state.[49, 50, 51]

9

Synovium is the delicate inner lining of the joint capsules (balloons) that surround most of the joints in your body. The inner lining of that balloon is covered with synovial cells that excrete synovial fluid, which is the primary lubricant of your joints. Synovium also lines the inner wall of the bursae of your body. Subacromial bursitis is a common finding in many shoulder problems. Well, the excellent results we get with problems such as bursitis or rheumatoid arthritis are directly because the synovium *loves the laser.*[52] This can help our rotator cuff example.

Therapeutic coherent lights are sometimes called cold lasers. There is no such thing as a coherent light/laser that makes cold light. This term has become popular because the clinicians want to make sure people know that their unfocused lasers are not warm, and they do not burn or harm. It is the same reason I have decided to use the term *coherent light* instead of *laser* interchangeably in this book. What else has the coherent light done for the *sick* cells in the chronic rotator cuff example? It has made those sick cells start *jogging*. It has been shown to **double or triple** cellular respiration.[53, 54]

It makes your cells start *"breathing deeply"* again and produce double or triple energy.[55] It also causes the release of growth factor[56, 57, 58] from illuminated cells. It rejuvenates the area. Coherent light also does many other proven things that I will cover in Chapter 5: Physiology of the Healing Cascade. The cells *wake up*! Then they try to heal again, one more time, with the start of this new healing cascade. The coherent light also stimulates new blood vessel formation![59, 60, 61] This is old news for those who study and use therapeutic coherent lights.

For the reasons above, and many other specific reasons that you will find throughout this book, people with chronic rotator cuff problems respond very well, especially when nothing else has worked.

- Coherent light therapy started in Northern Europe and Russia thirty years ago and has now spread throughout Europe, South Africa, Japan, Australia, Iran, Canada, and the United States of America. Russia actually has a Department of Laser Therapy, the "Laser Technology Centre of Russia."
- Much has been produced by Tiina Karu, the head of the Laboratory of Laser Biology and Medicine, Institute on Laser and Informatic Technologies of the Russian Academy of Science. I must say that this is one smart woman, and her work is so deep into biochemistry and molecular biology that I just do not know enough to understand much of it. She has been doing serious research that researchers around the world have respected for quite some time now. In her photographs, her cherubic smiling face tells me she is a very happy person.

- *This book operates on the principle of "low dose, low frequency of treatments with consistency." It is the reason for my belief that any normal person can safely be trusted to do this for him/herself. With each section, I inform the reader of the level of risk, so that you can make up your own mind.*

I expect that you will be aware and careful enough not to shine the coherent light in your eye, your child's eye, your pet's eye, or anyone's eye. I will also stress that you do not go beyond my warnings, particularly in the higher risk treatments, which I will identify. I will place targeted precautions throughout this book.

There are many concepts to coherent light therapy that I have "spread" throughout this book. For this reason, anyone planning to follow through on any of the following chapters is strongly advised to <u>read the entire book.</u> Come on, it's not that long.

All light in this dimensional universe is made up of photons. Coherent light is light in which there are packets of photons all organized and marching to the same beat. Every one has heard that focused lasers can blind you. Your store-bought red coherent light laser pointer will usually be a *near infrared* coherent light in the 630–670-nanometer (nm) range with a lens in front to concentrate the light into a *beam*. It is not *radiation* like x-rays. It is not ionizing radiation, which means that it does not cause skin cancer or anything like it. It has no ultraviolet light in it, unless someone specifically manufactures an ultraviolet coherent light. A red coherent light (like what I use in this book) cannot affect DNA. It is highly organized red light. A diamond is highly organized carbon. Although unrelated, each *fits* together in a cool way.

The discovery of the *bio-stimulating* effects of coherent light in 1967 was the result of a *researcher trying to cause cancer in mice with a laser*. It did not cause cancer, but it did make the shaved areas of the test mice *grow hair back at a much faster rate!*[62] Other researchers have tried to cause mutations with low-level laser as well but could not.[63]

You know, it is somewhat ironic that the discovery of bio-stimulation happened at a university named after Dr. Ignaz Semmelweis. Good medical care at that time meant leaving the midwife tradition and going to the hospital to deliver a baby.

The women were dying at his hospital from massive internal infections. He said something like, "Hey, let's wash our hands when we leave the autopsy room to deliver babies." The death rate dropped to less than 2%. In an interesting study of human behavior, he was fired, ridiculed, and driven out. This was partly because, in his zealousness, he was sending letters to local obstetricians, calling them murderers and such. He was eventually committed to a mental institution where, in two

11

weeks, the staff beat him to death. About 50 years later, they started washing their hands before delivering babies.

Dr. Endre Mester, Professor Emeritus of the Second Surgical Division discovered bio-stimulation with coherent light in Semmelweis University in Budapest, Hungary, in 1967[64]

Coherent light is just highly organized light. The only reason it is possible for coherent light to burn and cut is because they use a lens in front of the coherent light to focus the light to a pinpoint. I have just described the typical laser. Otherwise, the light would just spread out like a flashlight. I can take a magnifying glass and use the sun to burn paper with a concentrated pinpoint of sunlight. Without the magnifying glass, it does not happen.

Therapeutic coherent lights are intentionally built *without* a focusing lens so that they cannot build up heat. That is the reason why sizable therapeutic coherent lights do not have focusing lenses.

The therapeutic use of a coherent light does not cut, burn, or destroy anything. There have been over 2,400 research papers published over the past thirty years, and briefly, they prove that coherent light therapy is virtually harmless, even in much higher doses than I recommend. It appears that coherent light primarily has a significant effect on sick tissues and that it has little effect on healthy tissues. One way to describe it is that coherent light not only causes your sick cells to start *exercising*, but also to react with a healing cascade as if they have been injured. However, *no injury can be found*. All of those professional researchers, MDs, PhDs, and others, from universities and clinics all around the world, for the past thirty years *did not find harm* with much higher dosages than my recommended mild dose of 1–2 joules per area! (This works out to less than 0.3 joules/cm^2.)

As I have mentioned before, a number of specific published studies *try* to do harm with coherent light,[65, 66, 67] and it just does not do harm unless used in dosages hundreds and thousands of times larger than the small dosages I recommend in this book. On the other hand, numerous research papers point to a positive effect on surgery recovery when a therapeutic coherent light is used before or after surgery.[68, 69, 70, 71, 72, 73] (Healing time is sometimes cut in half, with a little more scar tissue than normal.) How much time exposure would give a dose of 2 joules of energy? A 100-milliwatt coherent light would give off 1 joule every ten seconds, so it would take twenty seconds to achieve a mild dose of 2 joules with the tip touching the skin, moving the coherent light slowly around the area treated (recommended at 3 square centimeters, which works out to 0.22 joules/cm^2). Using your 5-milliwatt laser pointer, you would need to illuminate an area for sixty seconds to achieve a dosage of 0.3 joules. If you keep the treatment area at 1 square centimeter, then you have a dose of 0.3 joules/cm².

A person should not use the coherent light on him/herself without some study, manual, or guidance, because, as an extreme example, if you are a weak/frail individual, you might use coherent light on or near the thyroid gland and cause a heart attack or stroke from a rapid heartbeat or rise in blood pressure that may follow. *Stay away from the thyroid gland, unless you are using our guidelines here.* *Even then, you are the ultimate judge of what you are experiencing. I will tell you when we are venturing into the unknown in a few of my chapters here, just so you remember to be aware! In addition, a little knowledge can be dangerous.* *Do not treat other people! It is illegal, and really, how do you know if they do not have an infection or cancer?* Leave that to the professionals. That goes for your children as well. *Do not use any of these therapies on a child!*

3

SAFETY[†]

GLAND AND ORGAN WARNINGS

An interesting study showed that the immune system *could* be stimulated with coherent light.[74] It also showed the *dangers of over-stimulation of the thymus gland.* The thymus gland is underneath your sternum (see figure 16 in Chapter 15.) They used 0.2 joules/cm^2, which is a low dose, near what we use here. Mice are much smaller, of course. They illuminated the thymus glands of mice *every other day for thirty days.* For the first five days, they saw an increase in immune activity. Then, for those mice that completed the thirty days, they ended up with *suppressed immune activity.* It could be said that *they burned out the thymus gland*, at least temporarily. If coherent light can cause exhaustion of one gland, it must be assumed that it can exhaust any gland.

Glands are critical components of your body. It is one thing to provide a small biostimulus; it is another to *burn it out* by repeated illumination with no rest.

How can you know? By starting with the guidelines in this book and tailoring them to your own unique body. Buy, and use a blood pressure cuff. Record your daily measurements on a calendar. Pay attention to how you feel. *Glands and organs must be treated more delicately than the rest of the body. If you notice <u>any</u> result, that is enough.* There is no need to increase the dose, as long as you are consistent with it. *In the case of glands or organs, you can know you have reached your "Maximum Medical Improvement" (MMI) when you notice <u>any</u> effects, even after only one session. That is when you <u>increase the time between treatments per the DelRe Protocols in Chapter 6, or as specifically directed in the particular chapter you are reading.</u>*

Remember our aim is to use less and less over time. I use a homeopathic approach in this book. The most any gland or organ should be illuminated is once a week, and that is only to get started. You want to cut back as soon as you have seen any positive effect. If you try to *whip* a gland or organ into an excited state, you will likely cause a negative outcome, one you may not be able to change back. You may then need to use hormonal therapy.

We still do not know much, so a person should stay within guidelines and know the things from which you should stay away.

WHAT NOT TO DO

1. *Keep coherent light out of eyes. Be aware of reflections. Keep children and pets out of the room.*
2. *Do not use if you may be pregnant. I do not know of any specific harm, but it is common sense not to treat a pregnant woman, particularly over the fetus.*
3. *Keep the laser away from the thyroid gland, unless you are following specific instructions. Even then, proceed with care.*
4. *Do not use over suspected cancer. (Research has not been able to show an increase in cancer growth in vivo,[75, 76, 77, 78] but it is common sense to keep the coherent light away from tumors.)*
5. *I do not know of any studies of laser therapy and endometriosis, but my guess would be that coherent light therapy would be contra indicated for endometriosis, as it may stimulate both cell number and robustness of the monthly sloughing off of the menstrual products trapped within the abdomen, as well as the inflammatory immune response.*
6. *People with light sensitivities should proceed with caution. Start with ¼ the dosage time. Some antidepressants increase light sensitivity, although I have not seen a hypersensitive reaction at this time.*
7. *People, who are extremely "frail," with very weak hearts or glands, should not treat the thyroid gland or the adrenal glands, and initial treatment should be limited to bedsores and superficial problems, such as bursitis, diabetic neuropathy, or muscular/tendinous problems.*
8. *Do not use a laser pointer that has a pointed tip. Use a laser pointer that has a flat circular tip.* Most $20 laser pointers bought at places in the U.S., such as Office Depot, Office Max, or Staples, now have ½" flat circular tips.

9. *If you have asthma, do not attempt to illuminate the lungs or bronchi. The bio-stimulation may cause a reaction and flare-up of your asthma.*
10. *Always keep the laser out of the reach of children. Always.*
11. *Do not treat organs or glands if you have a fever. You could cause the fever to go up. Something else may go wrong.*

Based on my experience, I have chosen to concentrate on chronic complaints and prevention and to avoid treatment of new injuries. Chronic complaints and prevention require small dosages and are therefore safer to treat. Glands require a small dose. All they need is a small stimulus.

Treatment of new injuries, although promising and useful, entails risk of too much scar tissue formation. Until more research comes in, a professional experienced in such work should handle these,

I believe it is good to take the complex and make it simple. There is so much variation in the research regarding dosages and protocols that it is like comparing apples to oranges. That is a problem. For example, many research papers say that the researchers used daily doses of coherent light. Based on my paradigm, that does not make sense. The paradigm I have accepted, and proven to my own satisfaction, is that *the most valuable result* of illumination of the coherent light is the response that stimulates a new *healing cascade* in sick cells. The healing cascade involves many molecular- and cellular-level synchronistic events that occur primarily over a forty-eight to seventy-two hour period, which culminate into a complex healing framework. Wouldn't daily coherent light sessions interrupt that amazing and complex process? I believe that they may be making the mistake of trying to *force* the body's cells and tissues to do something instead of *encouraging* the body's cells and tissues to do something. It is good to wake them up. It is a mistake to use a whip.

Other complex factors are the differing wavelengths and powers. There are also a number of *pulsing* therapeutic coherent lights on the market, with researchers comparing the efficacy of one frequency of pulses vs. another. I believe that as coherent light therapy evolves, we will see repeated proof that certain conditions or areas of the body respond better to particular wavelengths, treatment protocols, and pulse frequencies. However, for now, a person can safely and reliably use a standard 630–680-nanometer visible red laser pointer and achieve excellent results. It is a classic wavelength range in coherent light therapy. Even if the results can be further improved in the future with more specific protocols, most further improvement in the results would not likely be outrageously better, just better.

I am happy with what I do, using the coherent light to *wake up* sick cells that have *given up.* Because I work on musculoskeletal conditions, I usually limit treatments to twice per week, and when the patient has reached a peak or plateau (usually within four weeks), I cut back to weekly, then biweekly, then monthly treatments, and so on for chronic conditions such as fibromyalgia and rheumatoid arthritis.

Always remember safety. If there is any chance of children finding any coherent light, including the small 5-milliwatt laser pointers, the coherent light should be out of reach of children! Although the blink reflex will protect against eye exposures, with children, you never know what they might do to themselves, others, or even pets. Safety first!

For more information on coherent light research, you may want to visit www.laser.nu.

The lens that comes with your store-bought 5-milliwatt laser pointer will not cause a heat issue, because this FDA class of coherent lights (3A) is all limited to 5 milliwatts or less. Since the store-bought coherent light *does* come with a focusing lens, you *may* wonder why this is OK with me. Well, it is only 5 milliwatts. For perspective, it is *only* ½ of 1% of *just* one watt! That is almost nothing. This is one reason why the general field of medicine (and me for many years) ignored coherent light therapy. How can such a little bit of energy cause such a waterfall of effects? It is hard for doctors to believe, and even though it has been repeatedly proven for thirty years, nobody "important enough" seems to be paying attention. With enough of you telling them, they cannot ignore it forever.

When you experience success, you can order one or more books for friends and relatives, (we offer quantity discounts.) Please contact the publisher Health-1st via www.health-1st.com.

Right now, there is little real financial incentive for doctors to use coherent light therapy. The insurance companies will not pay for anything more than infrared therapy, which pays about $12–$20 per session. Nobody went to medical school to make that little money.

There is so much for mainstream medical doctors to know and study these days. The nature of the industry is that it gravitates to drugs, surgery, and specialization. For example, it would be difficult for an MD to make a living prescribing mostly nutritional supplements, because he/she is competing with everybody, including companies such as Herbalife, in that business.

It makes sense to study the things at which any doctor can make exclusive money. There is already more than enough for an MD to study within the areas of drugs and surgery alone. Physical therapists, on the other hand, would probably love to use coherent lights for these conditions, but they are captive to the prescriptions that come into their clinics. They rely on those doctors for income, so they cannot push too hard or insult them by "telling the doctor what to do," (which they should be able to.) That would be an aspect of *statistical process control* (SPC). SPC came from an American, but only the Japanese know what it is.

OK, back to safety. Even though your store-bought coherent light is only 5 milliwatts, you must be aware that if you were to physically hold your eyelid open so you could not blink and stare at the incoming coherent light beam from your coherent light pointer for a few constant seconds, you would experience eye damage, maybe even blindness in that short time. For perspective, you can also go blind by staring at the sun. In both cases, the same thing saves you—you blink.

This is the reality: if you happen to flash your eye, your body's natural blink reflex defense shuts your eyelids and makes it a non-issue, and *you would be just fine.* In other words, you automatically blink, and this happens so quickly that there is no harm.

Just remember: *Do not stare into the coherent light or any reflections of the coherent light from mirrors or glass, or you could suffer eye damage.* Your closed eyelid stops coherent light pointed in its direction (do not do this) but not so well, if you put the coherent light against your closed eyelid (do not do that either). This is pretty much the same advice as: don't stare at the sun.

Keep pets and children out of the room. You really should be alone, unless you need assistance. Perspective: these laser pointers have been used in classrooms and such for over fifteen years now. Kids buy them, and you or I have yet to see/ hear of any harmful incidents. People blink, and that is all it takes to protect you. *If you use a more powerful coherent light than 5 milliwatts, you should wear laser goggles specific to the wavelength of the coherent light you are using.* You can find them online at sites such as www.noirlaser.com, www.vs-medical.com, or www. wickedlasers.com.

There are plenty of warnings here, and I will be describing unproven and untested protocols and treatments as well. I will tell you what has been shown to be safe and what is likely, but unproven, to be safe if caution and discipline are used.

I will also cover treatments that are untested, that could entail risk, and that would likely cause negative effects if my guidance were not taken seriously. Warnings will be clear in those chapters. Do these prudently. You should inform your doctor about your activities. The odds are he knows very little about coherent light therapy. He may try to tell you not to try it, based on his lack of knowledge and scientific discipline of skepticism. Ask him if he/she will monitor your progress and tell you if anything looks wrong. If he does not want to, you will need to find a doctor who will monitor and, I hope, advise you.

This book has chapters on using coherent light on the thyroid and other glands.

19

Do not do these unless you have already bought an automatic blood pressure cuff. It will tell you how much of a reaction you are getting and if it is too much. Let me tell you directly that you will be experimenting on yourself in this area, so be careful to watch for negative side effects and discontinue self-treatment if negative effects occur.

4

CHOOSING A
COHERENT LIGHT†

At places such as Wal-Mart, you can buy a laser pointer for as little as $3. The only problem is that those laser pointers use small watch batteries, which will always need changing. In addition, they are often pointed, which is dangerous. Another issue is that the quality control is not that good, and you may get a weak one that only puts out 2 or 3 milliwatts. Most $20 laser pointers bought at places such as Office Depot, Office Max, or Staples now have ½" flat circular tips and are about 6" long, making them easier to handle, with lighter touch buttons. That makes it better if you are holding the button down for a few minutes at a time.

Those will use AAA batteries, which will last much longer than the watch batteries. Remember you only want the red color. Do not get the green ones, as that wavelength, though advertised as more powerful, cannot penetrate past the water, hemoglobin, and melanin in your tissues as deeply as the red. Speaking of dark skin, the darker your skin is, the more melanin you have. This decreases the amount of coherent light that gets through. My African-American patients respond as well as Caucasian ones, but I *do* use a 100-milliwatt invisible red coherent light for most problems, mainly because it is faster. Invisible red does penetrate melanin better than the visible red, though.

If you have very dark skin, you may need to increase the dosage time. I recommend you start with the dosages outlined in this book and perhaps increase the dose by 25% at a time until you get the response you are looking for. Doubling the dose should be the limit.

While in the store, try the laser pointers out if you can. Pick one that requires only a light pressure on the button switch. Look for the brightest red one. Remember, do not buy the green ones. I have seen articles that use green light because they consider it "cleansing." I believe this is silly logic. Yes, chlorophyll is classically considered a "blood cleanser." I believe it is because of its staining, thus binding, capacity in "capturing" impurities in the blood stream. However, even if you accept that proposition, if you think about it, as a color of light, green is the only one *rejected* by green things. You see it as green because it is the only wavelength that bounces off the surface of the green thing. So, why would green light be considered to be cleansing if green wavelengths are rejected, not absorbed by green things?

The above-mentioned laser pointers can keep their lenses, because 5 milliwatts is a very tiny amount of light energy. I believe the lens is helpful for increasing penetration depth for such a small wattage. As mentioned before, this is only one-half of one percent of one watt.

In the following chapters, I will provide my recommended dosages for a 5-milliwatt focused laser pointer, the highest power allowed for use by the public in the U.S. *This book is meant for an international audience and other professionals as well.* For that reason, I will give dosages for higher power coherent lights, which may or may not be legal to own in your country and may depend on your professional status. You should check your laws regarding laser usage. These lasers are available on the Internet from companies such as wickedlasers.com. *If it is legal for you to do so,* you can convert these types of lasers to therapeutic use with the following instructions:

1. Pry off the tip. (A butter knife works.)
2. Unscrew the lens. (I use the same butter knife.)
3. Put the tip back on while trapping one layer of plastic wrap as a clear dust shield.
4. Trim off any excess plastic wrap.
5. You now have a basic continuous wavelength therapeutic laser.

22

I believe I have covered the risks well in this book of *low-dose* therapy. It is my hope that this book will help fuel an explosion of interest in low-level laser therapy around the world, from the bottom up, and I like that. It is the essence of medical freedom. For doctors and therapists who are considering a more powerful therapeutic laser but do not want to pay for something that they are not sure of and seems expensive, I have provided the means to do so inexpensively right here.

To make the math easier, I chose 100 milliwatts for the higher power given in this book. I will always give my recommended dosage for a 100-milliwatts non-focused coherent light along with the 5-millliwatt laser pointer. If your laser has a milliwattage power between 5 milliwatts and 100 milliwatts, then you will want to change the dosage time accordingly. It is not necessary to use a more powerful laser in almost all cases. *Any laser pointer with a power higher than 5 milliwatts must have the lens removed!* This converts it to a therapeutic laser and greatly decreases the chance of eye damage or juvenile mischief. As always: Keep out of the reach of children!

The FDA has cleared the HairMax LaserComb Premium model. The premium model has nine lasers in it, however, it costs about $545.00 online. Some folks say it works for baldness; some do not. It probably depends on what type of baldness a person has and how far along it is. I am interested in including it in this book because nine 5-milliwatt lasers will cut your treatment time down to 11% of how long it would take with only one 5-milliwatt laser because you are producing 45 milliwatts total. You will not get the deeper penetration that a single 45-milliwatt coherent light would achieve. In addition, you may be treating bigger areas than you want to because of the size of the device. It would be good for treating big surface areas, such as bedsores, feet, forearms, shoulders, or bursitis of the hips. Not necessary, though.

I say that the important thing is to get started, so just go get a $20 laser pointer at Office Depot, Office Max, Staples, or other office supply house, and try it out.

As of the writing of this book, this is an excerpt from the FDA's Web site regarding laser pointers over 5 milliwatts:

CONSUMER SAFETY ALERT: INTERNET SALES OF LASER PRODUCTS

The Food and Drug Administration (FDA) is aware that some laser products being sold on the internet may not meet federal safety requirements and should not be available for purchase by the general public. In some cases, these products are overpowered and may be unsafe if not used responsibly. In other cases, these products are intended for use only by licensed medical professionals, trained operators, or other approved users.

FDA's CONCERNS

FDA is concerned about recent reports of laser products directed at aircraft—a potentially hazardous situation. The agency is particularly concerned about the increased availability of overpowered green laser pointers. Overpowered green laser pointers are those that may have been modified to emit more radiation than the manufacturer's original product.

WHAT FDA CAN DO

FDA's authority is over the manufacturers of laser products. These products must meet a federal standard for the amount of radiation they can emit and must be properly labeled. FDA is working to identify manufacturers of overpowered green laser pointers and other illegal laser products and will take action to prevent unsafe products from being sold in the United States. If illegal products are imported into the U.S., they may be refused entry, returned to the seller, or destroyed.

BUYER BEWARE

FDA recommends that consumers be cautious when buying laser products over the internet. Consumers may unknowingly purchase an illegal laser product or may lose their money if the illegal product is refused entry into the U.S. or destroyed.

Consumers should be aware that:

Medical lasers may only be sold to licensed medical practitioners.

Class IIIb and class IV laser light show projectors, identified as such on the label, may only be sold by or to individuals or firms with current, approved laser light show variances from FDA. Laser products that are advertised as uncertified components may only be sold to other manufacturers and may not be sold to the public for general use.

Laser products should have certification and identification labels (shown below) stating the product complies with the federal laser standard.

"Complies with 21 CFR 1040.10 and 1040.11"
"Manufactured or distributed by"
"Date of Manufacture "
"Complies with 21 CFR Chapter 1, Subchapter J "

Products should have a warning label advising the user to avoid exposure to the laser radiation.

Consumers who cannot verify the above, or do not understand what it means, probably should not sell or purchase the products.

5

PHYSIOLOGY OF THE HEALING CASCADE†

Here are a just a few research-derived attributes of various wavelengths and output power of coherent light on living tissue:

- Increased angiogenesis. (Increased blood vessel formation in wounds.)[79, 80, 81]

- Increased osteogenesis. (Increased bone formation in fractures, healing, etc.)[82, 83, 84, 85, 86, 87, 88, 89, 90, 91, 92]

- Decreased healing time. (A 50% decrease in healing time has been reported.)[93, 94, 95, 96, 97, 98, 99, 100, 101, 102, 103, 104, 105]

- Increased Human Growth Factor production.[106, 107]

- Increased ATP output (as high as 300%). This is another way of saying increased cellular respiration. (It is as if your cells are exercising aerobically.)[108, 109, 110, 111]

- Increased nerve healing and regeneration. (Severed nerves heal better when coherently lighted.)[112, 113, 114, 115]

- Healing of diabetic ulcers.[116, 117, 118, 119, 120]

- Treatment of obliterating atherosclerosis.[121, 122, 123, 124]

- Healing of arterial ischemia.[125]

- Increased local endorphin production.[126, 127]

- Decreased pain and tenderness of chronically inflamed tissues, including breast tenderness.[128]

- Positive results[†] with:
  - Fibromyalgia.[†, 129, 130, 131, 132, 133]
  - Rheumatoid Arthritis.[†, 134, 135, 136, 137, 138, 139]
  - Bursitis.[†, 140, 141]
  - Tendinitis.[†, 142, 143, 144, 145]
  - Prostatitis.[†, 146, 147, 148, 149, 150]
  - Heart Disease.[†, 151]

The effects listed above reflect the general effects of coherent light on *many types* of unhealthy cells in the body (in vivo) *and* in laboratory test tubes and such (in vitro).

Let's cover some of the events that take place when tissue is healing from a trauma or wound. There are two primary types of healing: *healing by regeneration and healing by repair.*

HEALING BY REGENERATION

Most tissues have this ability. Even when cells have died, and they have been removed by your scavenger cells, there is a framework, webbing if you will, of collagen left behind. Collagen is tough stuff. The remaining good cells regenerate by multiplying, and they build along the structure of the collagen web. The collagen webs are produced by specialized cells called *fibroblasts*. They function like spiders building a web, except fibroblasts are shaped like footballs with no legs. They lay straighter lines when exercise is applied. They leave a jumbled-up non-stretchable mess in a scar.

HEALING BY REPAIR

Some cells cannot regenerate. Cardiac (heart) muscle and your nerves cannot be replaced. Sometimes the collagen web has been destroyed. These areas heal by repair, involving *the healing cascade.* There are three phases of the healing cascade. In actuality, the healing cascade involves many different molecular and cellular actions. They must all happen chronologically, without interruption, to have a perfectly healed wound.

THE 3 PHASES OF THE HEALING CASCADE

1. Inflammatory Phase
2. Proliferative Phase
3. Maturation Phase

INFLAMMATORY PHASE (REACHES ITS PEAK WITHIN THE FIRST TWENTY-FOUR HOURS):

The body's scavenger cells are drawn to the area by the chemicals released by the injury and inflammation, and those scavenger cells generally consume the dead cells and bacteria and dissolve them in their multiple internal "stomachs." Fibroblasts and blood vessel-building cells are also attracted and begin their work.

PROLIFERATIVE PHASE (FUNCTIONING WITHIN THE FIRST FORTY-EIGHT HOURS):

Granulation tissue is formed as "filler" for the void left by the wound. It also is made by fibroblasts. It is supposed to be temporary filler, but in chronic conditions, it stays, becomes "sick," and does a relatively poor job compared to the original tissue.

You can think of granulation tissue as a "cheap grade" of tissue. With re-injury, repetitive stress, or interruptions in healing, you have more disorganized, lower-level granulation tissue and scarring.

Ordinarily, good original tissue cells that can reproduce do so and fill the area as well. With good care and light exercise for strains and such, the fibroblasts and other cells such as skeletal muscle cells have better parallel lines of stress to follow, and the resulting tissue is stronger, better, more organized, and healthier, with less internal/external scarring. As mentioned before, fibroblasts also make the stuff of scars when all else fails. The blood vessels in a chronically unhealed area of granulation tissue are also of a "cheap grade." They break and leak easily.

MATURATION PHASE (CONTINUES FOR TWELVE MONTHS OR MORE):

During the maturation phase, all of the "repair-crew" cells slowly leave the area. The tissue, with some scarring of varying quantities and qualities, will mature and remodel ("smooth out") over the next twelve months or more.

An unhealed rotator cuff injury of the shoulder has commonly "filled" with "cheap" granulation tissue. One way I was taught how to help heal a chronically unhealed rotator cuff injury was to re-injure it mildly in order to start a new healing cascade. It worked well for my patients about 75% of the time.

What I learned to do was to dig my finger into the front rotator cuff attachment and rub hard for a few seconds. (You can see the spot may become very red from the "cheap grade" of broken capillaries.) It was not pleasant for the patient by any means, and I did not like to do it, but only one treatment would be all that I would do. Usually, within a week, I heard from the patient that the shoulder was much better. Now, thanks to my coherent light, I do not need to subject my patients to such painful and potentially injurious treatment in order to start a new healing cascade.

Low-level coherent light has been shown to stimulate not only fibroblasts, but also blood vessel formation and other regenerative "repair-crew" cells, such as bone-building cells. (Refer to the beginning of this chapter.)

Some Russian coherent light researchers have stated that the reason cells respond so dramatically to low-level coherent light is that in all of evolution, our cells have never "experienced" coherent light. They are "surprised" by it, but *not harmed*. I will take that a step further and describe the response as a parallel to the healing cascade. The sick cells are surprised by "evolutionarily" unknown light, "think" they have been injured, and are stimulated to start a new healing cascade.

For me, coherent light is like a coach who rousts his team and gets them going or like a cheerleader who shouts to the team to try, just one more time. However, as in the case of glands mentioned in Chapter 3: Safety, if the coach is constantly working them too hard without adequate rest, the team can become harmfully "burned-out," and may never come back to peak performance. Please follow the guidelines in this book, and do not go beyond them.

6

THE DELRE PROTOCOLS†

The DelRe Protocols are a simple, safe, standardized way to take advantage of the bio-stimulating healing cascades initiated by coherent light therapy.

These protocols are for muscles, tendons, bursae, and most soft tissues. They *do not apply to treating glands or organs.* I will give specific protocols for glands and organs in the individual chapters on glands and organs. *Do not treat organs or glands if you have a fever.*

The DelRe Protocols allow for a maximum of 2 joules of energy to any 3 x 3 cm area included in this chapter. This provides 0.22 joules/cm². This gives a *mild dose.* If that area is cut in half, you will have a dose of 0.44 joules/cm². That is a *mild-moderate dose.* You do not have to know the math. I will give easy-to-follow instructions for you to follow in each chapter.

$$20 \text{ sec} \times 10 = 2 \text{ Joules}$$
$$60 \text{ sec} \times 100 \text{ mw} = 6 \text{ Joules}$$
$$60 \text{ sec} \times 5 \text{ mw} = 0.3 \text{ Joules}$$
$$7 \text{ min} \times 5 \text{ mw} = 2 \text{ Joules}.$$

 ## TREATMENT TIME AND AREA FOR A
5-MILLIWATT VISIBLE RED LASER POINTER

Use sixty seconds per 1 square cm area (the size of a small coin.) This provides 0.3 joules of energy/cm². The long time factor of 60 seconds has value, and this dosage *does* work. If you would want to reach the full 2 joules of energy I use with a 100-milliwatt coherent light, you would illuminate six spots within a 3 x 3 cm area for one minute each. The minimal dose of 0.3 joules/spot used consistently twice a week will work for most problems such as bursitis, bunions, rheumatoid arthritis, rotator cuff problems, tendinitis, etc.

29

The coherent light is held perpendicular, *directly against the skin*, with constant motion. For a 5- milliwatt laser pointer, you can hold it still for ten to thirty seconds on any one spot, then move to another spot within the area noted, continuing to move the light until the dosage time is reached.

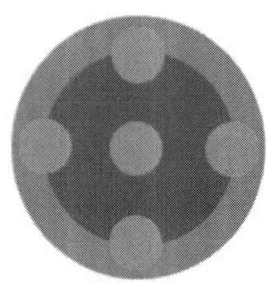

If you are treating a larger area, such as a rotator cuff, here is a representation of 5 doses to a 3 x 3 cm area

Treatment Time and Area for a 100-Milliwatt Visible Red Coherent Light

Use ten to twenty seconds per 2–3 square cm area. (This gives 1–2 total joules of energy.) The coherent light is held directly against the skin with constant slow motion or held still for three seconds on any one spot and moving to another spot nearby, continuing to move the light until the dosage time is reached.

Treat the areas twice a week or once every three days until you hit a "healing plateau." A healing plateau is reaching **"Maximum Medical Improvement (MMI)"** for now. At this point, you may be 100% better, 80% better, or some other number.

Note: This is not the true definition of MMI, because the insurance industry considers it the point at which you are as good as you are going to get, and they are done paying for your treatment. (However, we are not done quite yet!)

While we are defining things, there are two kinds of "negatives" in this book. The first type is a *negative study* or research result in which they simply did not find any significant results. The other type is a *negative response* to coherent light therapy. In this case, it means something undesirable happened, such as soreness from underlying osteoarthritis.

Let's say that you have improved to an extent but are not seeing any further improvement. Do not worry if you are not at 100%. Drug-free improvement at 80% is a very good thing. It is too soon to tell. It may mean that your chronic problem is going to take time. Usually, if a chronic condition is going to get better at all, it is going to take time, simply because it has taken so long to appear and has now been there so long.

Once you have reached MMI, there is no good reason to treat as often because your maximum treatment rate of twice-a-week is no longer "speeding up" healing. Now it is just going to take time if further improvement is going to be realized. At this point, you can double the time interval between treatments so that you are only treating it once per week. This may sound strange but so did homeopathy at first.

Therefore, to recap, you treat twice per week, until you reach MMI. You then treat your condition once per week. During your second month, many problems should be "gone." If not, you should see slower but steady improvement or at least stabilization. Cut back per the protocols below.

If, in the third month, you are still at your last MMI (even if it is 50%), then this may be the best you will get. You can then start cutting back on treatments further by doubling the time to two-week intervals (one treatment every two weeks) if you have not already.

According to the DelRe Protocols, if you have been able to *keep* your last MMI level by treating every two weeks, then after a month or so, you can go to three-week intervals for a couple of times, then four weeks for a couple of times. Then see how far you can go between treatments by adding one week at a time. It is common for people with fibromyalgia, diabetic neuropathy, or rheumatoid arthritis to stretch it to four to six weeks. This is outstanding for such an easy drug-free treatment. Some may find that they need to keep the treatments to three- or four-week intervals in order to stay at MMI.

Other conditions such as chronic rotator cuff problems may very well not need any further treatment after a just a month or so. Some conditions, such as bunions, may always need one treatment every four weeks. A bursitis problem may "go away" in a week, for good. Always try to treat less and less often.

So...

1. Start at two treatments per week until MMI (maximal medical improvement) is reached.
2. Decrease to once per week until MMI is reached.
3. Decrease to once every two weeks until MMI is reached.

If continued treatment is necessary:

4. Decrease to once every three weeks to keep MMI stable.
5. Decrease to once every four weeks to keep MMI stable.
6. Decrease to once every five weeks to keep MMI stable.
7. Decrease to once every six weeks to keep MMI stable.

"Flare-ups"

If, after your self-treatment, you notice anything negative, such as an aggravation of pain, it may mean a number of things:

- You may be affecting an osteoarthritic joint. It may be underneath the tissues you are treating. Except for the area where the back of the skull meets the neck, you should stay away from the spine because of the possibility of "flaring up" some hidden osteoarthritis there. I make a distinction when it comes to bulging disks later in the book.

- You may be naturally more photo-sensitive than others are or because of certain conditions, or it may be drugs you are taking. I recommend that you cut the dosage to 25% of the dosages above, and give it a week between treatments. If that still aggravates your condition, you should stop treatment. If your condition improves, then you may slowly increase the dosage to the normal protocols listed above, if possible. Sometimes, it just takes a while for the body to get accustomed to the coherent light.

- People with rheumatoid arthritis (RA) can have simple wear-and-tear arthritis also. I must say, for some reason, people with RA seem to be "immune" to arthritic flare-ups from coherent light. You may be an exception if you experience the aggravation typical of osteoarthritis. You may not be able to benefit from this treatment. I have not yet seen that happen.

7
THE SHOULDER†

Rotator cuff injuries can be very difficult to treat successfully. Early on in my career, I did not use active rehabilitation in my office. When I received a patient with a rotator cuff problem, I would advise that he or she see a physical therapist for it. Later, when I added active exercise rehabilitation to my office, I had moderate success with the cases that had rotator cuff problems. *That* was before I started using coherent light therapy. Not only have I experienced surprising success with failed rotator cuff surgery cases, *they no longer need do any specific rotator cuff exercises.* I realize that statement is enough to make many chiropractors and physical therapists cringe, and it does go against logic and conventional wisdom. After all, the "cure" is to exercise the rotator cuff muscles so that they do their job, pull the shoulder "ball" back into the socket, and properly hold it there.

I have only seen two failed rotator cuff surgery cases. We were able to get both cases just about pain-free in thirty days. The other non-surgical cases have responded in kind.

A couple of patients have reported a minor aggravation of their shoulder the next day or so following treatment. In those cases, I believe that there is some osteoarthritis present, even if not significantly demonstrable on x-ray. Osteoarthritis is simple wear and tear of a joint. If a person has significant osteoarthritis in more than a few joints, then genetics and lifestyle are involved. As I have previously said, although some research indicates that coherent light therapy can help osteoarthritis, the only relief I have seen is with very small doses. It is very easy to "flare up" an osteoarthritic joint with coherent light therapy.

I do have many of my patients perform general body stretches and exercises. The exercises augment/enhance a patient's recovery. I do not consider them necessary when it comes to problems that I am treating with the coherent light.

Impingement Syndrome/Subacromial Bursitis/ Supraspinatus Tendinitis

(High Confidence/Low Risk)

This first treatment chapter is about shoulder treatment because I have been so delighted with the results of coherent light therapy on difficult rotator cuff problems. I will start with impingement syndrome/supraspinatus tendinitis, *the most common forms of shoulder bursitis/tendinitis.* It is usually present with rotator cuff syndrome.

Most people do not know that this is a cause of their shoulder problems, so check yourself now:

Find the flat-like bony plate at the *tip* of the shoulder (See figure 1.A).

1. Slide your finger from (A) toward the back of your neck about 1–2 cm, finding the tip of a soft "V" shaped depression (B). Press and see if it is painful, and probe to determine how painful it is. If it is painful, *this is the spot to treat* with your coherent light.

2. Probe back to the *tip* of the shoulder, (A). If this is painful, you may also have subacromial bursitis. (See figure 2.) You can use the coherent light right on a painful area of bone as well.

3. Refer to the DelRe Protocols for my standard treatment protocols (Chapter 6).

Figure 1: Shoulder, Superior View

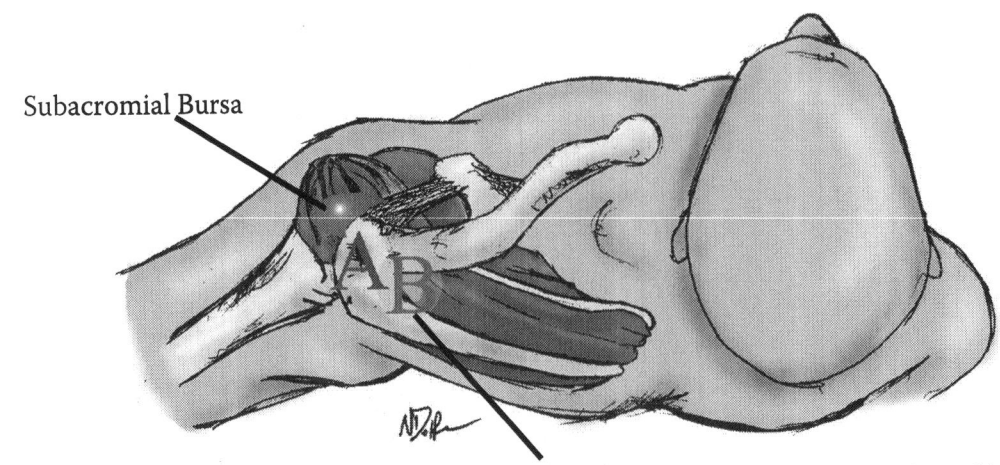

Probe/press "B" to look for tenderness. If so, this is indicative of supraspinatus tendonitis/impingement syndrome.

Subacromial Bursa

This is where you press your shoulder in order to see if it has a painful area. This is also the area you will treat. This is the "front" of the rotator cuff.

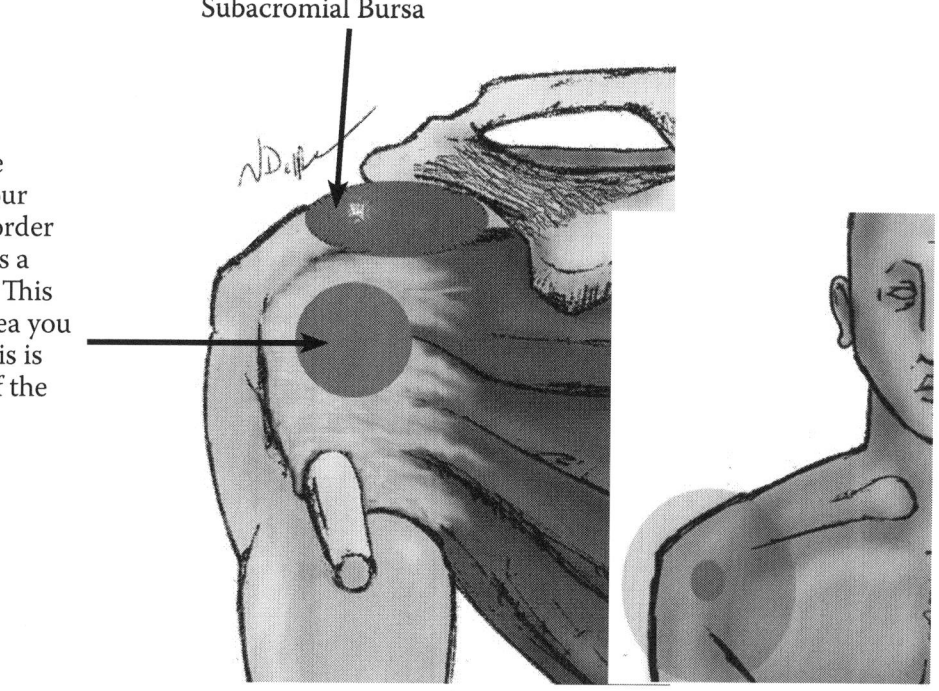

Figure 2: Shoulder, Rotator Cuff

This is the subscapularis muscle. It lines the inside of your shoulder blade, and is one of four muscles that insert into the rotator cuff.

ROTATOR CUFF SYNDROME[†]

(HIGH CONFIDENCE/LOW RISK)

Turn the affected arm outward, with elbow bent, like a hitchhiker. Probe the area with the purple spot in the front of the shoulder, shown in figure 2, with your other hand. It will be rounded like a ball. This is the front of the rotator cuff.

If it is painful/tender, treat it. Illuminate the area, using a flat-tipped 5-milliwatt red laser pointer against the skin. You want to treat four or five 1-cm spots to fill a total area that is about 3 x 3 cm. Please refer to the DelRe Protocols in Chapter 6.

8

Bursitis†

ALL TREATMENTS COVERED IN THIS CHAPTER ARE HIGH CONFIDENCE/LOW RISK.

The best way to describe a bursa is to imagine you have a grape between your thumb and first finger. You squeeze one side of the grape so that the fruit pops out of it. You are still holding the peel between your fingers. Now, if you lightly rub your fingers together with the peel in between them, you will feel a slippery feeling from the inner lining of the grape peel rubbing against itself. That is how a bursa works. When the body has two tissues rubbing against each other, the "Technology of God" often puts a self-lubricating bursa in between those surfaces to avoid irritation from the rubbing. That is what a bursa does, and that is what a bursa is.

Bursas "love the laser". If you have been told you have bursitis, and if the diagnosis is correct, you are in luck! You have a great answer here in your hands.

Common types of Bursitis:

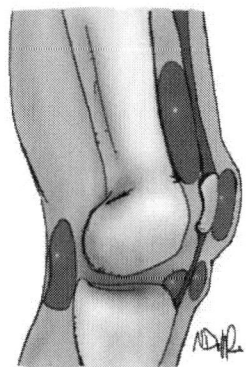

Knee: Prepatellar bursitis: often called "housemaid's knee" where continuous kneeling on hard surfaces inflames the **prepatellar bursa**, which is between the kneecap and the skin. Other local bursas are shown as well, in red. *Remember that coherent light can aggravate underlying osteoarthritis that may exist, so just stick to the main bursa.*

Figure 3: Knee, Side View with Bursae

Shoulder: Subacromial bursitis: Often "wrapped up" with supraspinatus tendinitis and impingement syndrome. (See Chapter Seven.)

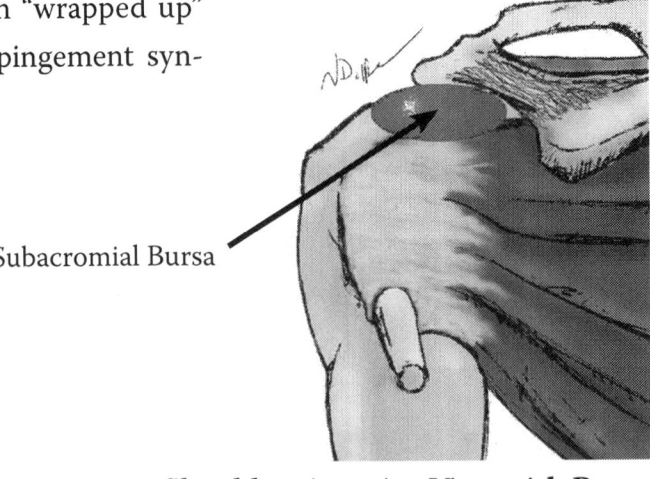

Subacromial Bursa

Figure 4: Shoulder, Anterior View with Bursa

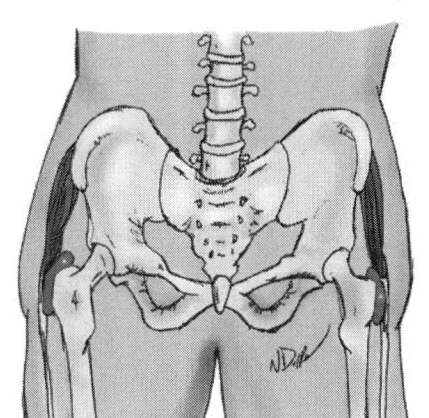

Figure 5: Pelvis, Anterior View

Hip: Trochanteric bursitis: Stand up, hands at sides. If you press in against your hips, each hand should touch the bony protrusion of your "hip bone". That is where that bursa is. A quick, easy way to find out if it is bursitis is to "pinch" a good two to three centimeters of tissue at that point, against the bone. If that is where your pain is, then you have found bursitis, and you have the answer to *__easily__* healing it right here in your hands.

Bunion: A bunion is more than bursitis alone, but the bursitis is the major cause of pain. You can stop it, starting today.

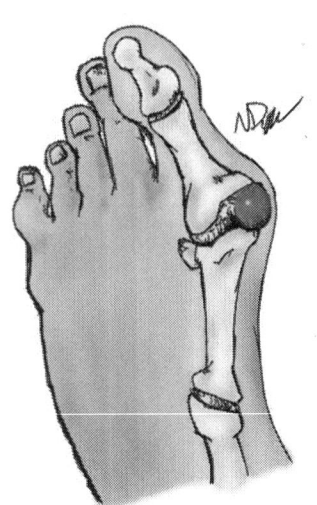

Figure 6: Bunion

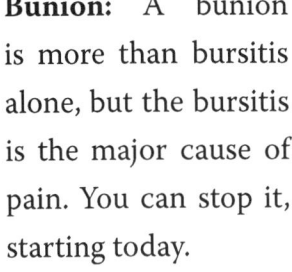

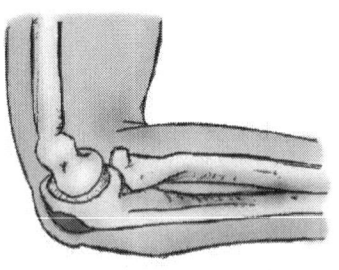

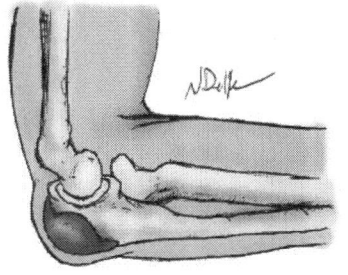

Elbow: Olecranon bursitis: often called "student's elbow", is swelling at the tip of the elbow.

Figure 7: Elbow

To treat all the above bursitis examples, plus any other bursitis you are diagnosed with, please refer to the DelRe Protocols (Chapter 6.)

9
RHEUMATOID ARTHRITIS†

TREATMENTS COVERED IN THIS CHAPTER ARE HIGH CONFIDENCE/LOW RISK.

Well, if you have rheumatoid arthritis, I probably have the first good news you have had in some time: You are luckier than those people with plain old osteoarthritis are. Don't believe me? Well, all you need to do is get a laser pointer and prove me wrong.

The reason you are luckier is that coherent light does very little for osteoarthritis (simple wear-and-tear arthritis.) In fact, the coherent light will typically aggravate that condition, with little, if any, benefit later.

You, on the other hand, will experience quite the opposite. You will experience more mobility right away. You will have a lessening of bad days and an increase in good days. Most of you will continue to improve over a month or two and will achieve a level of painlessness that you had given up on long ago. It is significant and very important to stress that this is not like using drugs. This requires less and less treatment as you improve. You can possibly go six weeks between treatments, or longer. It is a blessing for me to be able to give you this good news!

Always start with the metacarpal-phalangial joints. What? Well, look at figure 8.

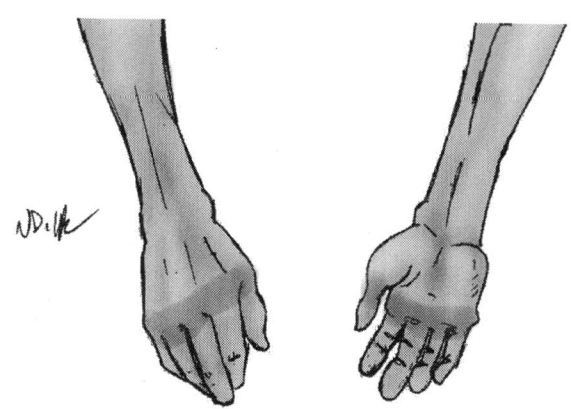

Figure 8: Hands

Scan the purple bands on both hands for four minutes per surface, for a total of sixteen minutes, with a 5-milliwatt laser pointer. If you are using a 100-milliwatt coherent light, use twenty seconds on each of the four surfaces. After you have treated the hands, you may treat other troublesome joints using the DelRe Protocols in Chapter 6.

As mentioned in Chapter 5, there is a "systemic effect" with rheumatoid arthritis. In other words, simply by treating some joints, *all* the joints have a positive response in the patient with rheumatoid arthritis. I believe that this is because of the positive effects of coherent light on the blood and the tissues it supplies. See the specific chapters on the joints you want to treat if they are available in this book. For example, if you are going to treat the shoulder, please review Chapter 7. If you have any wrist or forearm problems, you should read Chapter 11.

If you experience any negative effects, such as an aggravation of a certain joint, then skip that one for a while. If this happens with more joints, you may not be a good candidate for this treatment. Try cutting back to 25% of the dosage and dropping back to one treatment per week, treating only the joints that do not flare up and improve. Your system may be able to take more later.

I just realized that *it took two pages* here to cover rheumatoid arthritis. Take a moment and think of the tons of steroids such as prednisone and the chemotherapy drug methotrexate that are consumed around the world by these unfortunate people.

From what I have read, and what I have seen with my own eyes, if you use my information here, you may be able to put down most of those drugs for good. No more calcium stealing, immune-system ruining, organ-destroying prednisone.

Remember President Kennedy? He was a skinny kid when he went into office. Then, we synthesized cortisone. Next, his face looked like a chipmunk because he was taking too much for his back pain. If he had survived long enough, he would have been riddled with osteoporosis and arthritis, as well as a laundry list of other ills. Sometimes good medical care can be made much better.

10
DIABETIC NEUROPATHY†

Diabetes "loves the laser." I know I have already used that expression, but it also does apply here.

Regarding the direct treatment of the pancreas with coherent light in order to improve its health and insulin production, I do not believe that exposing the surface of the body to coherent light can penetrate deeply enough to directly affect the pancreas. Considering that there are indeed effects, *the blood* appears to be carrying the bio-stimulating effects to the pancreas. The pancreas is mostly hidden behind the stomach and small intestine. In addition, the pancreas is not close to the back of the body either. (The kidneys/adrenals are close to the back.) The depth depends on the person, but I would say the surface of the pancreas is at least 6.3 cm (2½ inches) deep. A 100-milliwatt visible red coherent light has an effective depth of 4 cm, so theoretically, it should not have a direct effect.

Research, however, shows benefits. In fact, just one research project used intravenous 2-milliwatt coherent light blood therapy along with 5–20-milliwatt treatment outside of the liver, spleen, and pancreas. They did this daily for eight days on 218 patients. They did this again at three months and again at six months. Sugar-decreasing medications were decreased 200%. (I do not know how you can decrease something 200 %.) Body and extremity warmth increased, and blood fats and cholesterol decreased. [152] Although this is intriguing, I am not going to come up with any protocols for such treatment regarding the above subject in this book, because it is not certain how to do it properly, and these people can be very sick, with little room for error.

However, we can treat any diabetic's foot numbness with confidence, and many diabetics will find that many of the other physical problems they have are also listed in this book. They generally respond very well to coherent light therapy with any of the problems covered in this book. You can use the same protocols for other areas of poor circulation due to diabetes.[153,154,155]

The pancreas is illustrated here as the little spot behind and below the stomach. Other organs hide it.

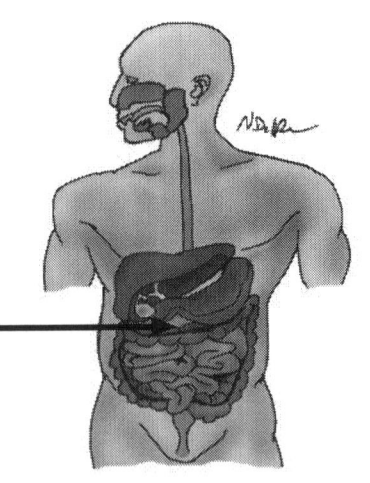

Pancreas

Figure 9: Abdomen

FOR DIABETIC NEUROPATHY:

(HIGH CONFIDENCE/LOW RISK)

Treat the areas over the nerves shown in Figure 10:

Instructions for a 5-milliwatt laser pointer: five minutes on each surface. Move the laser   pointer slowly along the course of the nerves from the toes to the instep and back again. Cover the surface of the top of the foot for five minutes and then the bottom of the foot for five minutes from the toes through the arch. It will take twenty minutes to do all four surfaces on both feet. Start doing this twice a week, then cut back per the DelRe Protocols in Chapter 6. If you get any pain in your toe joints the next day, it is the osteoarthritis, so just try stay to away from the joints next time. If using a 100-milliwatt coherent light, treat each surface for twenty seconds.

If you have diabetes, you should be taking *grape seed extract* daily. You do not have to do this to make coherent light therapy work, but you should do it anyway. Grape seed extract can heal a detached retina (diabetic retinopathy), and if it can heal it, it can prevent it. Grape seed extract helps to compensate for the poor microcirculation found in diabetic people. The highly active bioflavonoids

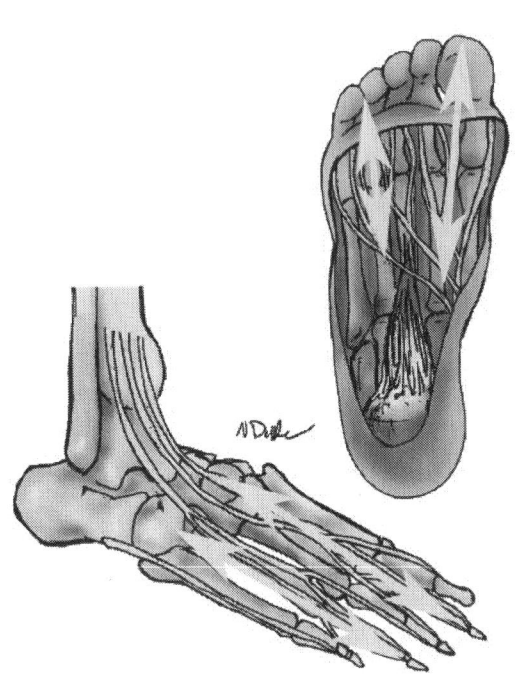

Figure 10: Foot with Nerves Shown

are what your body needs to form the microscopic "bedding" that your capillaries live in, providing protection, nourishment, and health for these tiniest of blood vessels. You need it. If nothing else, drink green tea every day, as green tea has bioflavonoids similar to grape seed extract. I have actually seen grape seed extract help macular degeneration on one person (and just about *nothing* works on that). The dosage is 20 mg for every 20 pounds of body weight (or 100–300 mg daily) for three to four weeks reducing to a 40–80 mg daily maintenance dose.

The more physical exercise you have, the better. If you are diabetic, there is no avoiding the obvious. Do something, anything.

One more tip—I have repeatedly seen diabetics' blood sugar (and medication dose) drop as much as 30%–50% with the addition of 1–2 ounces of liquid "colloidal" minerals to their daily routine. The best ones I have used are found at www.newvision.com, and at 1-800-MINERALS. At half the price, good liquid minerals ("Liquid Naturals") used to be available at Sam's Club. The web site is www.remingtonhealth.com. I have recently tried another type at: http://www.vitalearth.org. It is called "**Fulvic Mineral Complex**". The last one named has a chelating agent, so you may go through a period of detoxification as it pulls heavy metals out of your body.

An expected drop in blood sugar means that you must be constantly checking your blood sugar and adjusting your blood sugar medication, so consultation with your doctor should be considered mandatory.

11
CARPAL TUNNEL SYNDROME†

TREATMENTS COVERED IN THIS CHAPTER ARE HIGH CONFIDENCE/LOW RISK.

In 1994, General Motors took part in a study to see if low-level coherent light therapy really did help carpal tunnel syndrome.[156] In seven plants, the company experienced an 85% success rate. The results were so striking that the FDA formed a completely new category for this type of treatment. Research has repeatedly shown that coherent light is safe and effective for carpal tunnel syndrome.[157,158,159]

Carpal tunnel syndrome is experienced as pain or numbness radiating from the palm-side of the wrist, extending into the hand or arm, or both. There is a space/tunnel in the wrist that the median nerve passes through, and it can get trapped/squeezed there. The General Motors study consisted of using coherent light on the wrist itself.

My approach involves not only the wrist, but also the forearm. Please review figure 11.

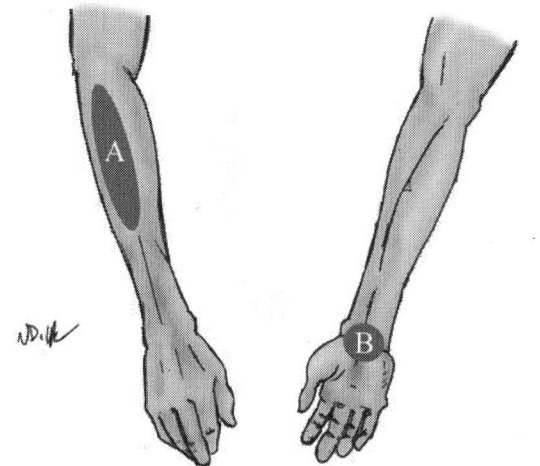

Figure 11: Forearm and Wrist

Palpating means that you are probing with your fingers. Place your hand and forearm on a table, palm-side down.

Your free hand will be the palpating hand. Feel for the long bones on the sides of the wrist. (Squeeze side to side.)

After you find them, follow them up the forearm, about halfway.

Now, *change your palpating hand position, rotating it* so that you are squeezing the forearm *muscles* that are *between* those *bones*, and squeeze through the arm between the two bones. (Figure 11.A.)

Probe and squeeze the muscles, looking for areas of tenderness, between the two long bones of the arm, working your way towards the elbow. Actually, the massage you are inadvertently doing is beneficial to carpal tunnel syndrome. If you find something "ropey" and painful, you have found it. If your middle finger pops up, you are a "hot one." This will work great on you, and your carpal tunnel will improve.

DEEP MASSAGE OF THE FOREARMS

This is an important thing to do, in addition to the coherent light therapy. Do this in the beginning, no more than once per week. I have found this to be quite reliable for clearing out carpal tunnel syndrome, even before I started using coherent light therapy.

I believe that when there is myofascitis of the forearm musculature (it is sore and stiff when squeezed), it torques the two arm bones you have found like a couple of chopsticks and pulls on the carpal bones as well.

When that happens, the eight tiny carpal bones of the wrist (that form the carpal tunnel arch) "flatten out" that arch, just a bit, and the tunnel becomes smaller.

When you relieve those forearm muscles of their abnormal tension, with massage and coherent light therapy, those two arm bones return to their proper relationship, and the carpal tunnel arch is restored, thus relieving the nerve pressure. I am trying to keep this simple, but if you are curious, the long bones are called the radius and ulna.

Therefore, the coherent light therapy to the *forearm* is meant for treating the sick, sore muscle and fascia, and the coherent light therapy applied to the *wrist* is meant to help heal the sick squeezed nerve found in the carpal tunnel.

HOW TO TREAT CARPAL TUNNEL SYNDROME

48

Illuminate the area of forearm pain (figure 11.A) using the DelRe Protocols in Chapter 6.

Illuminate the palm side of the wrist (figure 11.B) also following the DelRe Protocols.

Many, many carpal tunnel cases have their root cause as a stuck section of the spine, between the shoulder blades, and up towards the neck. If your carpal tunnel problem does not resolve with the above instructions, or if it keeps returning, then you should seek a recommendation for a good chiropractor in your area. You should not opt for surgery until you do the above recommendations.

12

MALE HEALTH†

Research has shown that coherent light therapy can help produce more testosterone,[160] healthier and more numerous sperm[161], and can heal a sick prostate gland.[162,163,164,165,166,167] Is that enough?

We are venturing into less secure territory here. *This is a higher risk chapter* because glands are very different from other tissues, and glands are very different from each other.

The *effect coherent light has on glands lasts for about a week.* In addition, while it is considered conventional wisdom that coherent light only has a significant effect on sick, morbid tissues, this does not appear to apply to glands.

Glands appear to be stimulated whether they are sick or healthy. In fact, healthy glands may react *more* than unhealthy ones. The glandular system acts like a "liquid nervous system." It sends more than just impulses, though. It switches elements of glands on and off. The brain switches most of the glands of the body on and off primarily by way of the pituitary and, to a lesser extent, the pineal gland. The concentration of the various contents of the blood switches them on and off. Please see Chapter 20 for more details on the complexity involved.

There is much we do not know, so I must warn you that there may be unwanted side effects, or long-term effects, that we do not know of. *Let me tell you that you will be experimenting on yourself when it comes to glandular therapy involving coherent light.* I am revealing this information so that you have more choices available, not because I am advising that you do these things. All I can say is that I have used many of these techniques on myself with no adverse effects. I can also say that, so far, in the thousands of research papers that exist, there have been no adverse side effects with the dosages and frequencies of treatments that I recommend in this book; however, long-term studies regarding the information in this chapter are not available.

Do not treat organs or glands if you have a fever. You could cause the fever to go up. With a fever, the body is already in a hyper-stimulated state.

THE PROSTATE GLAND†

(HIGH CONFIDENCE/MODERATE RISK)

Does it help the prostate? Certainly. Regarding the testosterone and sperm-enhancing effects mentioned above,[159] I just do not know for sure. Prostatitis used to rear its ugly head occasionally, and I have I treated it successfully according to the method below. Prior to that, I successfully used one tablespoon of vinegar in a glass of water before bed for a few nights. If you decide to try these glandular methods, depending on the chapter, I advise that you do it in consultation with a specialist in that discipline. Regarding this chapter, you must first rule out cancer, infection, sexually transmitted disease, and a host of other things that may be the problem. If the diagnosis comes back as plain prostatitis, then you can safely illuminate it with coherent light.

Coherent light has been shown to cause the immune system to act against cancer when there are only the first few bad cells (dysplasia),[168,169] so this treatment may very well prevent prostate cancer. That is just an educated guess, and not supported by specific research.

Although the research regarding coherent light therapy to the prostate involves insertion of the coherent light into the anus, you do not need to do that. *Insertion is unsafe and unnecessary.* Because the blood carries the effects as well, you can treat the prostate through the skin. The easiest way to find the right "spot" is to stand up, reach down between your legs, and find the bone that is just in front of the anus. (The cross-section shown in figure 12 does not show that "bone"; actually, it is the lower portion of the pubic symphysis.) You want to treat between that "bone" and the anus. (There is not much room there.)

You also want to use thin plastic wrap to cover the laser to keep it sani-

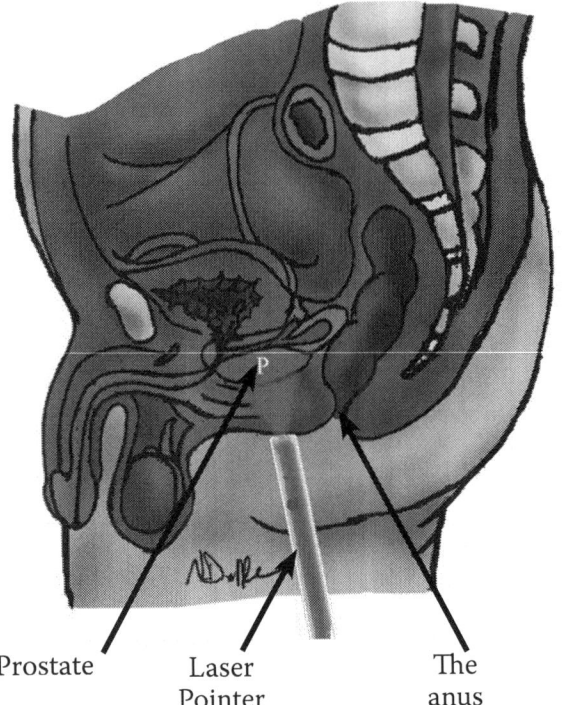

Prostate   Laser   The
         Pointer   anus

Figure 12: Prostate Gland Illumination

tary and wash your hands afterwards. You can treat it lying on your back so that that area can relax and the laser can be pressed a bit deeper into the tissues towards the prostate. (See figure 12.P.)

Do not use a laser that has a pointed or rounded tip. Most $20 laser pointers bought at places such as Office Depot, Office Max, or Staples have ½" flat circular tips. These are what you want. They use two AAA batteries.

Do not press too hard or otherwise hurt yourself.

You want to point the front of the laser pointer straight up, a little forward, **not** backward towards the anus). (See figure 12.P.)

Treat that area for three minutes with a 5-milliwatt laser pointer. I know this breaks my "sixty seconds in one cm² spot" rule, but there is just not much room there around which to move it. *If you are light sensitive*, your skin may become irritated for a day or so. Decrease the dosage time next time by 50%, and see. If you are using a 100-milliwatt therapeutic coherent light, you will treat this area for twenty seconds.

When you are illuminating this area, you are also illuminating the "root chakra." This is considered a primal "power point" in Ayurvedic traditions. You may feel a boost in energy. For illumination of glands or organs, consider *any response to be MMI.* A *"response"* can be when you feel better or notice any improvement. This can also include a negative response. If so, you need to stop or cut back on dosage time, depending on the severity of the reaction, whatever it may be. When you have *any* response, go to the next lower frequency of treatments:

1. Start at once per week until *"response"* is reached.
2. Decrease to once every two weeks until *"response"* is reached.
3. Decrease to once every three weeks to keep *"response"* stable.
4. Decrease to once every four weeks to keep *"response"* stable.
5. Decrease to once every five weeks to keep *"response"* stable.
6. Decrease to once every six weeks to keep *"response"* stable.

51

If you have Peyronie's disease (Peyronie's disease is an abnormal curvature of the penis caused by scar tissue in the erectile tissue), you can illuminate the entire penis, spending more time on the scar tissue. With a 5-milliwatt laser pointer, illuminate the penis for three minutes. With a 100-milliwatt coherent light, you would treat for twenty seconds. Research backs this up.

The conclusion of those researchers? "We consider therapy with noninvasive laser to be absolutely the most effective component of the whole therapeutical complement within the scale of possible therapies of Peyronie's disease."[170]

TESTOSTERONE[†]

Testosterone is not a nice hormone. The prisons are filled with men who produce more testosterone than normal. These are just a couple of the observations of a prison testosterone study:

"The higher the level of testosterone, the more gruesome the crime. Lower levels of testosterone were consistent with non-violent crimes. In addition they discovered that the inmates who had higher levels of testosterone had violated more prison rules and were serving extended sentences for crimes during incarceration. The study was repeated in a **woman's correctional** *institution and although the overall testosterone levels were lower the ratio of increased aggression to testosterone was the same with the higher levels signaling more violent crimes and lower levels were associated with non violent crimes."*[171]

Think of those old black-and-white movie clips of a fist-pounding Hitler, jerking around, giving a raging, frenzied speech. Now, think about the fact that German chemists had just synthesized testosterone, and he was taking it. Apparently, a lot of it. Makes one wonder.

Synthetic testosterone was new. Coherent light therapy is thirty years old.

Once again, you should consult a specialist in these matters before proceeding with your own therapy in order to rule out other diseases and monitor your progress. I have tried this for my own research purposes. When I did, I noticed the same feeling as when I had tested testosterone-building supplements on myself in the past. I took them for a month. I became very jumpy, anxious, and aggressive. My blood pressure was up. That was enough for me, so I stopped.

With that in mind, men's testosterone levels generally decrease, as they get older, depending on a number of factors. As a result, they tend to gain weight, lose muscle mass and strength, and have a decrease in sex drive. It is common for them to have testosterone patches prescribed by their doctors. They may have another option. It appears from the research that coherent light therapy to the testicles can increase testosterone production.[172]

THE TESTICLES[†]

(MODERATE CONFIDENCE/HIGH RISK)

With a flat-tipped 5-milliwatt red laser pointer, treat each testicle for thirty seconds, constantly moving the laser pointer. Cover the entire surface of the testicle, all around it.

With a 100-milliwatt unit, illuminate each testicle for five seconds. If you experience any tenderness or ache the next day, cut back to half the dose next time.

Start at one treatment every two weeks, and cut back as testosterone levels rise, which may be with the first treatment.

How do you know? Well, along with blood tests with your doctor, *you need to own an automatic blood pressure cuff.* You should know what your blood pressure should be, and it should be under control. If your blood pressure rise is one to nine points for a few days, I believe that is acceptable. I am talking about either the upper (systolic) or lower (diastolic) numbers. If your blood pressure rises higher than ten points or stays high for more than three days, you should wait two weeks and cut the dose in half.

The nervous system and the "liquid nervous system" of the glands both have positive and negative feedback loops. These are similar to electronic circuits. The brain "sees" "enough" testosterone in the blood. It shuts off the liquid switch called follicle-stimulating hormone (FSH). (FSH is the hormone that turns the testicles "on.") Therefore, in theory, it turns the testicles "off." If it is because of using *steroids* for sports, the testicles may shrink. Why bother then? Well, the aim is to rejuvenate the testicles, not to inject artificial testosterone into the body. When this is done slowly and prudently, the body will hopefully be kept "in the loop," and things should balance out. I suspect that the sensors that "judge" the "right" levels of testosterone will adjust to expect a higher level. I mean, if the feedback mechanism were perfect, then the threshold levels would not go down in the later years. I believe that the threshold *standard* may adjust to match the health of the testicles.

Treatments should be no more than twice a month to start and *take it to four weeks when you notice <u>any difference</u>, such as intolerance, anger, anxiety, aggressiveness, or achiness in the testicles.* Your tests ordered through your doctor would be an objective way to measure your results. You can also order testosterone saliva tests online. Be aware of the *blood pressure response.* It is discussed in more detail in later chapters. Any rise in blood pressure is a signal to cut back to once a month between treatments.

Use as little as possible to get an effect. Remember the research that implies that *you can burn out a thymus gland, if it is treated too often.*[173] You do not want to burn out your testicles. *Remember, this is a <u>high-risk</u> procedure because of the unknown risks.* I am not telling you to do this; I am simply revealing this information to you. If you choose to try this treatment on yourself, use the schedule of decreasing treatments displayed earlier in this chapter in the prostate section.

What are the long-term effects? It may be that a man could form a dependency and require testosterone patches. The opposite may happen with hyperactivity of the testicles. I suggest you avoid getting a woman pregnant if you are actively treating the testicles. It may affect sperm in some ways that we do not know. What we

53

do know is that coherent light will increase the amount of sperm produced and their level of activity. If you and your spouse are trying to conceive a baby, and you have a low sperm count, you may want to use this method.

Because of the unknown, you should wait to try to "make a baby" for at least one month after your *last* treatment. In addition, you should be taking 50 mg of zinc a day for the health and number of sperm and for the prostate. It is said that 70% of a man's zinc is stored in the prostate. In addition, simply said, a young man growing up zinc-deficient will have a smaller penis than his twin who had enough zinc.[174]

The reason I feel it is important to tell you these things is because of the tiny amount of power involved. A physicist would tell you that this could have *virtually no physical effect* on the atoms and molecules that make up your cells. The small dose I recommend of thirty seconds per testicle with a 5-milliwatt laser pointer amounts to 0.15 joules of energy. *That is only 1/16 of a calorie! That is the primary reason I consider showing people how to do these things. The dose is extremely small. It is the same energy as a tiny 1-watt light bulb that is on for ⅓ of a second.*

In the case of this book, that energy is in the form of visible coherent light. Although you can be out in the sun, you cannot stare at the sun, can you? The same goes with coherent light, particularly when it has a lens that forms a laser beam. It is not "radiation" in the sense of high-frequency wavelengths such as ultraviolet or x-rays, which *can* have an affect on physical matter. The only physical effect it could have is to slightly warm the tissues (like a tiny red heat lamp), but that would take much more than a 100-milliwatt unfocused coherent light to do this.

The **physiological effect** leaves us with the benefits but also the unknown risk. It is the thirty years and 2,400 published research papers that decreases that unknown risk. That is good.

We may eventually find that, instead of building a dependency, the opposite may happen in some glands—*they may become more active and stay that way* when treated for a number of months. Remember, there are many negative side effects to high testosterone levels. You do not want a stroke, do you? *So...be careful. Be aware. Be adult. Consult with your physician. Pay attention.* I expect the glandular protocols in this book to be changed, as we know more.

13
FEMALE HEALTH†

THE BREAST†

(HIGH CONFIDENCE/LOW RISK)

Research has shown that chronic inflammation increases the risk of cancer in many tissues.[175,176,177] There is also research that specifically demonstrates that chronic breast inflammation can increase the risk of breast cancer.[178] Chronic inflammation and tenderness is an invitation for something to go wrong.

Coherent light therapy has been shown to decrease pain, inflammation, and size of benign fibrotic lumps in the breast.[179] Coherent light therapy stops breast tenderness and inflammation.[180,181,182] I cannot say it more clearly.

From my wife's experience, one treatment will yield results (the day after treatment is usually pain-free, and it stays that way for months). You may need more treatments, so start at one treatment per week, then cut back according to the DelRe Protocols in Chapter 6. Go to two-week intervals quickly, as the breasts have glands in them. Get to once per month as soon as you can. Less is always better.

Some women actually have their breasts removed before they get cancer, because it runs so strongly in their family. Would coherent light therapy help those women? Only time will tell. If it were my wife, I would advise coherent light therapy and regular check-ups including mammograms.

Do not treat organs or glands if you have a fever.

Assuming your physician has examined you:

With a 5-milliwatt laser pointer, you will spend three to five minutes on each breast, depending on size. Concentrate on the most tender spots and fibroids for about thirty seconds, but otherwise, cover the entire breast, lower and upper. Press in with the *flat-tipped* laser pointer at deeper tender structures *without injuring yourself.*

With a 100-milliwatt coherent light, you would illuminate each breast for twenty seconds, stopping at tender spots for two to three seconds.

Endometriosis[†]

I believe it is wise to *stay away from using coherent light therapy on endometriosis*, as it may stimulate both cell number and robustness of the monthly sloughing off of the menstrual products within the spaces between the intestines and may do the same thing to the inflammatory immune response. These are what contribute to the pain of endometriosis.

Premenstrual Syndrome (PMS) Cramping[†]

(Low Confidence/Medium Risk)

Chiropractic spinal adjustments have been shown to help those with PMS.[183] If you do not know why, please do a search online, or see a chiropractor.

Some chiropractors give nutritional recommendations for PMS in order to help decrease the back pain it may be causing. One aspect of PMS is cramps. Calcium/magnesium helps with most muscle cramping, including cramping associated with PMS.

Because of my license restrictions, I cannot use the following coherent light instructions to treat this problem in my office with coherent light therapy. However, if my wife had that problem, I would suggest she try, just to see if it helps. Remember, muscle "loves the laser," and the uterus *is* a big muscle (with a very specialized inner lining), so I would suggest treating the uterus with coherent light therapy *between* menstrual periods.

Make sure you do not get pregnant during this because of the unknown risks. Do not use coherent light on the uterus during menstruation. Stay away from the ovaries. Why? Because they are just too important. You are born with all the eggs you will ever have. Even with the conservative dosages in this book, I would not suggest going near the ovaries. We will have to wait for more research to come in.

How do you know where the ovaries are? They are roughly halfway up between the pubic bone and the navel, and then both to the left and right of that point by 2–3 inches. (See figure 13.) Just *stay in the midline of the body.* That is where your uterus *is*, and that is where your ovaries *are not.* Wait until your menstrual cycle has stopped for a few days and things have settled down.

56

Treat it at *once per week until you notice __any__ difference*. Make sure you stop treatment at least a week before your next cycle/symptoms typically start. Therefore, you will probably have a treatment window of about two weeks. That would give time for two treatments between periods. If the treatment does help with your cramps, then cut back to one treatment the next month and see.

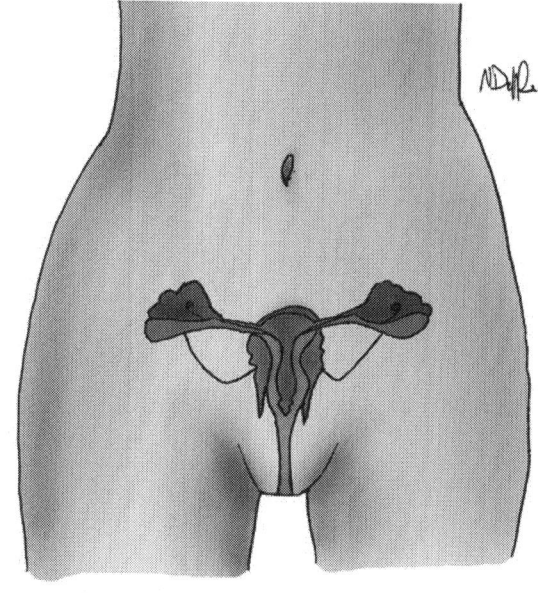

Figure 13: Female Anterior View: Uterus and Ovaries

With a *flat-tipped 5-milliwatt laser pointer*, lie on your back and bend your knees, feet on the floor, so that your abdomen is relaxed. *Start at the top of your pubic bone, and slowly work your way up until you are halfway to your belly button.*

Press in without injuring yourself, every half inch or so, for about thirty seconds per spot. Depending on your size, you should spend three to four minutes doing this. If you reach the halfway point before your time is up, slowly go back down, and so on.

14

FIBROMYALGIA† 纖維肌痛症

TREATMENTS COVERED IN THIS CHAPTER ARE MEDIUM CONFIDENCE/LOW RISK.

After about eight years in practice, I had given up trying to treat patients with what I call "full-blown" fibromyalgia. I would give them a phone number for a support group, tell them to try to insure that they got good sleep, and wish them luck.

Why? Because I had already tried the recommended supplements and treatments, and they just would not get much better. A soft massage and *maybe* light exercise was about the best physical care for these folks. If I were to adjust their spines, I would only be able to do it once or twice, and that is all. Otherwise, they would flare up from the treatment. Now, with coherent light therapy, I can help most, but not all, get an average of 75 % pain relief. Considering the alternatives, that is tremendous.

When I found that these patients did very well with coherent light therapy, I consulted with some researchers about it. One researcher, who held both MD and PhD degrees, was a man who worked for drug companies trying to get their drugs approved by the FDA. He said his group had recently failed at getting a drug approved for the treatment of fibromyalgia. He said that at that time, the FDA had not approved any drugs or devices for fibromyalgia. The reason drug companies were having trouble getting approval was because a realistic etiology had not yet been established.

What does that mean? It means that they could not prove what was causing fibromyalgia, and the FDA's argument was that if you cannot prove the cause, how could you provide a treatment? That *is* the essence and necessary discipline of medical science.

There has now been a drug approved for the treatment of fibromyalgia—Lyrica. One of the classic descriptions for fibromyalgia has been "rheumatism of the muscles," and this has been the view all these years. However, Lyrica does not treat muscles.

PAINFUL DISORDER.

It targets the nerves. It is a "nerve drug." This is especially interesting because the only other effective drugs that unofficially help fibromyalgia are the antidepressants. They also have an effect on the center of the nervous system—the brain. In a sentence, they help the brain to build up levels of serotonin, the "feel good" hormone.

So why do we have an epidemic of people, particularly women, who have fibromyalgia? What is happening to their nerves? Why does it seem to happen to people with healthy lifestyles? Why is it that the ones who "don't take care of themselves" do not seem to get fibromyalgia? Why is it that a "muscle problem" is really a "nerve problem"?

It is almost as if nerve toxins are causing a reaction in the nervous systems of these people with otherwise healthy lifestyles. They eat more vegetables than the average person does, and that should be healthy, right?

Pesticides can mimic nerve poisons. There are many researchers taking a closer look at these facts at this time. I have said that this book is based on science, but it is not scientific. I do take liberties. That means that I am not making any commitments to the following conjecture. That said, here are the long-term effects of the WMD sarin nerve gas:

"Long-term health effects of sarin exposure that have been reported include a decrease in *memory* in emergency rescue personnel three years after the Tokyo subway attack."[188]

Eye symptoms, *fatigue, muscle stiffness, and headache*, and psychological symptoms in survivors of the Tokyo subway attack five years later.[189]

"Animal studies have found that subclinical doses of sarin impair T cell responses and **decrease corticosterone** levels (Kalra et al., 2002) and result in alterations in muscarinic receptor sites in the brain that may be associated with memory loss and cognitive dysfunction." [190]

(I find it interesting that there is a decrease in the body's natural anti-inflammatory agents. That invites inflammation.)

REGARDING GULF WAR SYNDROME:

"The participants reported multiple common medical symptoms, including affective problems (50%), fatigue (42%), joint and *muscle aches* (40%), cognitive problems (26%), *headaches* (26%), respiratory complaints (24%), *gastrointestinal problems* (22%), *sleep disturbances* (21%), and skin problems (19%)."[191]

Since nerve gases can be described as "a molecule away" from pesticides, is there a link, and what can we do about it? Vegetables are used to "clean the system," but they may also be a cause because of pesticides. Nerves take a long, long time to heal, if they are going to heal at all. I honestly believe that the "Technology of God" will heal, or heal "around," most problems in time if we can remove the aggravating factors.

I had an insect exterminator at my house recently. I asked him if any of the "old timers" complained of muscle pain. He said "no." You know those fellows *have* to be breathing a certain amount. I know *he* was not wearing any mask, so who knows?

ANTIDEPRESSANTS[†]

Antidepressants *do* significantly help fibromyalgia.[192] What do you think the chances are that a large percentage of people with Gulf War Syndrome are on anti-depressants? If you would rather use natural supplements, 5-HTP is sold in health food stores. It aims for the same basic result as the antidepressants. It increases serotonin levels. It does this by supplying the building blocks of serotonin (the "feel good/anti-anxiety" hormone). SSRI Antidepressants work differently by stopping the brain from destroying serotonin, thus you get a build-up. If you are going to try out 5-HTP, start with 50 mg per day, and if that does not seem to do enough to help you sleep at night or relax, the dosage can be increased by 50 mg at a time. The maximum dose is 300 mg a day, which I think is very high. If you are taking too much, you will become more anxious. 50–100 mg per day will do it for most folks. If you are already on an antidepressant, you need to be warned about a theoretical risk of producing too much serotonin in your body. This would be accompanied by an increase in anxiety, heart palpitations, and in a few extreme cases, it has caused death. It is called "serotonin syndrome." There are no known cases where this has happened to someone taking 5-HTP, but it is something you should know about.

Now, on to the coherent light treatment protocols for fibromyalgia. I have a protocol for all fibromyalgia patients, and I treat what I call "the ring of fire." In this, I start with one forearm, and then I treat the muscles at the top of both the shoulders, then the other forearm. (See figure 14.) I then treat other muscles as needed. I use the protocols in Chapter 6.

THE "RING OF FIRE"[†]

Why do I choose these areas? I always treat the "ring of fire" because other than the upper spine, that is where fibromyalgia always starts. I do not use coherent light therapy on the spine because it may flare up an underlying osteoarthritis. In my experience, most fibromyalgia starts with a stuck, dysfunctional area between

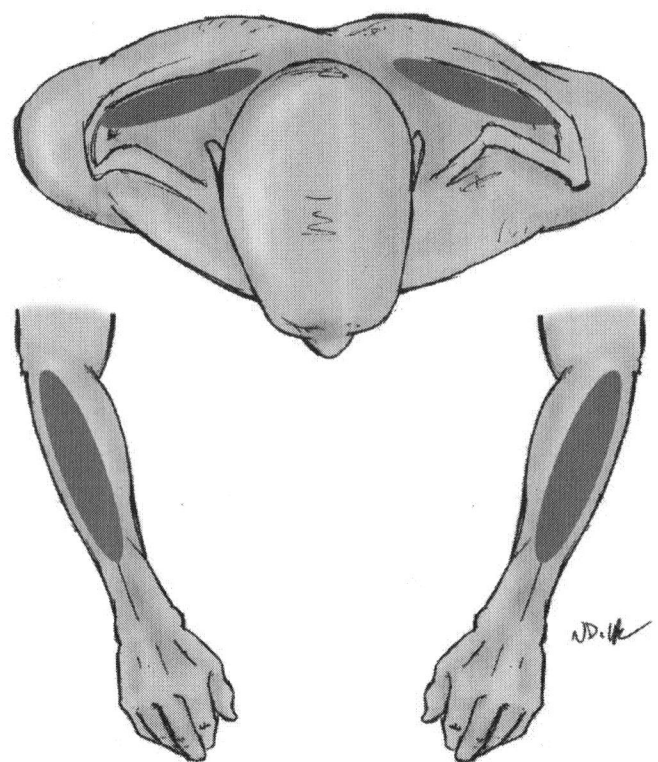

Figure 14: The Ring of Fire

the shoulder blades and the neck. This is called a subluxation. A subluxation can adversely effect local blood and lymph circulation, and have adverse effects at the nerve root, and along its course. It can cause dysfunction in the vertebrae above and below, by way of architectural compensation.

If the person has been consulting with most health professionals, he/she has not been sent to see a chiropractor. A professional will rarely send you to a chiropractor, although these days they often say nice things when you tell them you see a chiropractor. It is someone who cares about you that sends you to his/her chiropractor.

If this goes unchecked long enough, month after month, year after year, "muscle pain" spreads into the upper shoulders, the neck, and down the arms. It can eventually become *systemic*, and all muscles are painful. This is what I call "full-blown" fibromyalgia.

For a diagnosis of fibromyalgia according to the American College of Rheumatology guidelines, a person must have pain at eleven or more of the eighteen "special" points. With what I call "systemic fibromyalgia," you can poke one hundred points, and they all hurt. I simply squeeze the major muscles, the arms, the thighs, calves, etc. If the patient recoils with all four limbs and tells me that he/she still feels me squeezing after they pull away, they have systemic fibromyalgia. Many doctors believe they have little choice but to eventually give them opioids regularly. Pain pills. Can someone tell me how many humans can take opioids over a long period and not become addicts? These people are often left helpless. When they become addicted, they sometimes find it difficult to get their doctor to keep giving them prescriptions because now they are addicts. As I understand it, heroin is cheaper, and I have seen a case when that is where they go next. I am glad I am a chiropractor and do not have to make those prescription choices.

There have been no good, significant answers if the antidepressants do not work. (They, to some extent, usually do, thank goodness.) Now we have Lyrica. It has been reported that it does work; however, the pain is worse if they stop. Perhaps they may be stopping too soon, I do not know. I will not go into the side effects here, but you just *know* that a nerve drug is going to have a list of serious side effects. Will it be worth it? Only time will tell.

On the other hand, coherent light therapy has no negative side effects, is drug-free, and generally knocks down that fibromyalgia by 70%. The only problem is that there is no money in it, and nobody cares to see. You have to do it yourself. In addition, it does not cure it; it is a treatment.

You may be asking, "I thought he said that pesticides are the problem." I am just saying that pesticides deserve a close look. What I can tell you is what I have seen for over eighteen years in practice. Most fibromyalgia starts out as described above, in my experience. Now, with a typical chiropractic patient, they may simply show the signs and symptoms of local pain and perhaps some *nerve-pinching* signs, such as numbness or tingling down an arm. In most of those cases, fibromyalgia is not involved. Numbness and tingling may be part of severe fibromyalgia, but fibromyalgia is not usually a part of the typical patient with those same signs or symptoms. Perhaps sensitivity to pesticides makes the difference. One other factor that would single out women may be the fact that certain shampoos and conditioners have known nerve toxins in them.[193]

Look at this article written Tuesday, January 11, 2005, by Mike Adams of naturalnews.com:

63

Researchers at the National Institutes of Health have found a correlation between an ingredient found in shampoos and nervous system damage. The experiments were conducted with the brain cells of rats and they show that contact with this ingredient called methylisothiazoline, or MIT, causes neurological damage. Which products contain this chemical compound MIT? Head and Shoulders, Suave, Clairol and Pantene Hair Conditioner all contain this ingredient. Researchers are concerned that exposure to this chemical by pregnant women could put their fetus at risk for abnormal brain development. In other people, exposure could also be a factor in the development of Alzheimer's disease and other nervous system disorders.

My mother once said that over the years she noticed that men who used Head and Shoulders shampoo saw their hairlines recede. At http://www.baldingblog.com/2005/12/14/head_shoulders_shampoo, you can see entries by 126 people, most of who are complaining about Head and Shoulders shampoo making their hair fall out. In those anecdotal reports, many say that some hair grew back when they stopped. I added the above quote to help them with some more information, so if you go there, mine will be entry 127. A post followed it from an M.D., doing his job to question the information. I checked further, and added this: "At: http://www.naturalnews.com/003210.html he says this: 'These finding were presented December 5th at the American Society for Cell Biology annual meeting.'" That will be entry 166.

15
HEART AND BLOOD[†]

Low-level coherent light has a profoundly positive effect on the "sick" heart. [1] [94,195,196,197,198,199,200,201,202,203,204] What about the long-term effects of continuing such treatment? The longest study on coherent light therapy to the sick heart I have found was a 10-year study with 280 patients involved. They demonstrated many significant positive results and did not report any negative side effects.[205]

MY STORY

So, what is my experience? A number of years ago, I was noticing angina on a daily basis, so I started taking an aspirin a day, and that took care of it. My diet was not a good one. One of the most successful diet programs to heal hearts is the Ornish program. The best way to say it in one sentence is this: The closer you are to a vegetarian diet, the healthier your arteries will become. Meditation helps. OK, that was two sentences. Anyway, my wife is a wonderful cook, but we get few fresh vegetables in our meals. (What can I say? I can't get her to change.) I am not shirking blame; I love fast food, too. Yes, I know better, but I just enjoy it. And it is fast.

I know that in the end, a near-vegetarian diet is what the sick heart needs. With any nutritional approach, consistency is the key. Too many people find the right answer, but they are sporadic in their application of it. Please refer to Chapter 18 for important calcium/magnesium information related to the heart.

Back to my story: After a year of taking aspirin to ward off angina, I was in my office alone one day. I was starting to get a little angina again, even though I was taking the aspirin. I was thinking about a friend my age who was getting his *second* angioplasty/stent inserted right then, in that hour, in a local hospital. I did not yet know about the coherent light heart research that had been done. I said to myself,

"Oh, what the heck, the heart is a big muscle, and muscle 'loves the laser.'" So, I took my shirt off and gave five mild–moderate doses to my heart. I did it by putting my 100-milliwatt invisible red laser over my heart, and pressing against the skin, I slowly followed the spaces between the ribs over my heart in order to get the most penetration (about 4 cm, or 1½ inches).

I gave one twenty-second dose per rib-space, and I illuminated along five rib-spaces. Then, I went back to work. About an hour or so later, I felt warmth in my chest, and the best I can describe it, is that it was "happy warmth." You know, the feeling you get when someone "warms your heart." Another way to describe it is as an internal knot untying, a relaxation of sorts in the heart area. I have accomplished the same thing with a 5-milliwatt red laser pointer.

My angina went away for about six months. When it came back, I treated it again. After a couple of years of this, I found that I had to treat my heart more frequently. Eventually, I was treating my heart every four weeks or less. That went against everything I have seen with coherent light therapy. The goal is to need less and less of it over time, not more and more. That is the model to stay true to in this book. It is a homeopathic approach.

Since my outcomes did not match the model, I knew I had to do something else.

Therefore, I did what I had to do; I started eating right. I am a regular at Subway now because I eat on the run, and that is the only non-salad fast food I know of that has a bunch of fresh vegetables. I also am a regular at McDonald's, but now, I buy the salads. I never thought I would be satisfied with a salad, but they include a nice sized slice of chicken breast, and I stay satisfied. I promise they are not paying me to say that. I am sure the other places are good, too. I stopped eating as much of my wife's cooking, and I eat many more vegetables at home.

The outcome? *My angina went away within a month.* It has not come back. It probably will someday, and then I guess I will have to start exercising, of all things. (I hope the reader will recognize my version of humor when the spirit moves me.) In the meantime, I have already shown that the coherent light makes cells "exercise" and increase cellular respiration as much as three-fold. That will do for now. There are people who do eat right and still have heart problems. This chapter is for you.

These are my two approaches for treating the heart. One is *direct*, and it is the method I just described, directly illuminating the heart between the ribs. The other is *indirect*, via the blood stream leading to the heart. You see, the jugular veins in the neck are very close to the surface of the skin, and that blood goes directly to the heart. Treatment of the blood with light has been done for eighty years. There is probably even more research into illumination of blood with monochromatic light than there is for coherent light therapy.

The Russians have been busy with this since the early 1930s or so. Pure light of one color can have profound effects on the blood, which in turn, improves a host of diseases. The effects of non-laser illumination of the blood has been repeatedly documented. Particularly interesting was the research into exposing a liter or so of blood to ultraviolet light, then re-injecting it into patients with various ailments. The results were the most promising method for treating infections at that time before antibiotics came on the scene. Interest in blood illumination receded after that.

Since the advent of lasers in 1960 and the discovery of laser bio-stimulation in 1967, you just *know* the people familiar with blood illumination therapy would jump on this. Therefore, of course, most of this "coherent light on blood" research also comes from Russia.

So, what does it do? Well, first, remember that we are dealing with something much safer than ultraviolet light. Ultraviolet light is ionizing, which means it can knock electrons off your DNA, thus can cause skin cancer. The red laser pointer is simply producing red light, which has photons marching in step. For some reason, this really "surprises" any cells in the body, but the biggest changes come from the "sick and tired" cells.

COHERENT LIGHT THERAPY OF THE BLOOD[†]

COHERENT LIGHT THERAPY OF THE BLOOD HAS BEEN SHOWN TO:

Increase red blood cell number.[†, 206]

Increase the amount of the important antioxidant SOD.[†, 207]

Decrease angina.[†, 208]

Increase blood flow in the microcirculation.[†, 209]

Make platelets more slippery.[†, 210,211]

Decrease cholesterol levels, while increasing the good HDL cholesterol.[†, 212,213]

Improve heart rhythm on EKGs.[†, 214]

Improve blood circulation and the strength and flexibility of red blood cells.[†, 215,216,217]

Strengthen the immune system.[†, 218]

Well, if that was not enough, I must say I believe that somehow the illuminated blood carries with it the bio-stimulating effects of coherent light therapy. Why? Because of the research done with ultraviolet light and because of the reaction of patients with rheumatoid arthritis. Coherent light has a definite *systemic* effect on rheumatoid arthritis. In other words, you treat one hand, and the other hand gets better too. The blood is carrying something, and it is not found in the above research; it is something else. One recent study shows that when you add even a small amount of recently illuminated blood to untreated blood, it hands over the bio-stimulating properties to the untreated blood[219] (like one Olympian runner passing the torch to another). What is it carrying?

As more research is done regarding this, I predict that we will see that illuminated blood carries the bio-stimulating effects to not only other blood cells, but it also carries bio-stimulation to the body's organs and glands as well. In other words, we may find that it is like injecting "liquid coherent light" into the bloodstream. I see this effect, and I am going to capitalize on this hypothesized effect in this and some other chapters.

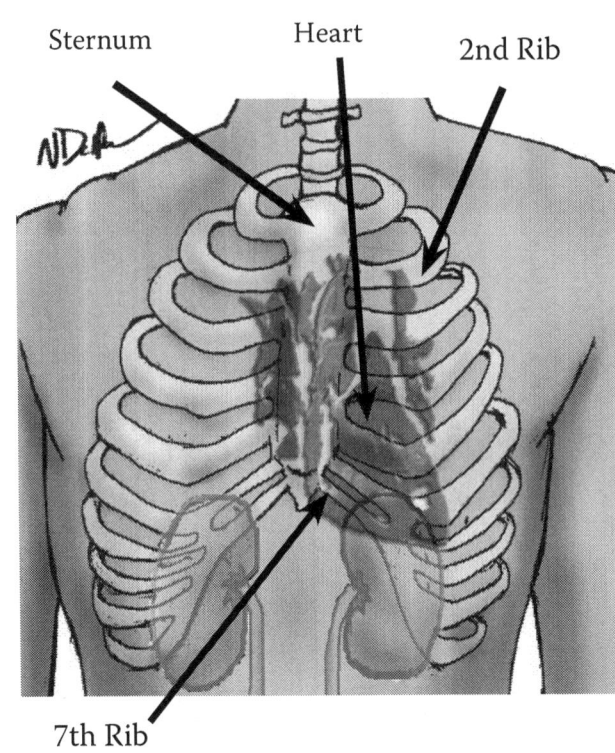

Figure 15: The Heart

Do you see why I used the word "Rejuvenation" for the title of this book?

DIRECT TREATMENT OF THE HEART[†]

(HIGH CONFIDENCE/LOW RISK)

Please refer to figure 15. You can see that part of the heart is under the center "bone" called the sternum. Note that the rest of the heart is on the left side of the chest under the second to the seventh ribs. Lightly pressing a *5-milliwatt laser pointer* into the skin, you want to spend two minutes slowly running *from the left side of the sternum, along one long space <u>between</u> the ribs.* Go as far left as halfway above the left nipple. As you illuminate the lower rib-spaces, go further left. Please refer to figure 15.

Start at the space between the second and third ribs. Do not be overly concerned if you are not sure. This is a safe procedure, and there is room for error, *as long as you do not have asthma.* You want to spend two minutes slowly going back and forth. Then, do the same at the next level down (rib-space between the third and fourth ribs). Do this for each level, until you have treated five of the rib-spaces.

If you press into your sternum and find tender spots, treat those. If you are using a 100-milliwatt coherent light, use twenty seconds per rib-space.

For illumination of glands or organs, consider *any response to be MMI.* The positive response we are looking for here is a feeling of relaxation in the area of the heart or decreased angina. Heart tests, such as EKGs, should improve. When you have *any* response, go to the next lower frequency of treatments:

1. Start at once per week until *any response* is reached, four weeks at the most.
2. Decrease to once every two weeks until *any response* is reached.
3. Decrease to once every three weeks to keep *the response stable.*
4. Decrease to once every four weeks to keep *the response* stable.
5. Decrease to once every five weeks to keep *the response* stable.
6. Decrease to once every six weeks to keep *the response* stable.

If you have *asthma,* you should *not* illuminate the lungs or bronchi. It could cause an extreme "reaction" and flare-up of your asthma. Since there is almost no way to treat this area without affecting the lungs, I believe you should not do it if you have severe asthma. If you have mild asthma, tread carefully. Start with ¼-dosage time, and work your way up to normal dosages if you have no problems. Otherwise, you should just stop.

The thymus gland is under the sternum. It is the main immune system gland but shrinks, as we get older. A very nice study showed that the immune system *could* be stimulated with coherent light.[220] It also showed the *dangers of over-stimulation.* They used 0.2 joules/cm^2, which is a low dose, near what we use here. We are one thousand times bigger than a mouse, though. They illuminated the thymus glands of mice *every other day for thirty days.* For the first five days, they saw an increase in immune activity. Then, for those mice that completed the thirty days, they ended up with *suppressed immune activity.*

It could be hypothesized that they "burned out" the thymus gland, at least temporarily. *Remember our aim is to use less and less over time.* The most any gland or organ should be illuminated is once a week, and that is only to get started. You want to cut back as soon as you have seen any positive or negative effect.

Do not treat organs or glands if you have a fever. You could cause the fever to go up. With a fever, the body is already in a hyper-stimulated state.

69

Indirect Treatment of the Heart and Brain†

(High Confidence/High Risk)

Please refer to figure 16. Probe your neck with your fingers. Find the pulse on both the left and the right. These are the areas to illuminate.

This could be a higher-risk procedure if you have a very frail heart, uncontrolled high blood pressure, or you have an over-active thyroid gland (hyperthyroidism). There are other unknown risks. If in doubt, you should not do it. If your heart is not well, but you can walk around OK and still want to try, start with a ¼ dose.

The reason I feel that it is safe enough to discuss openly is because this book is grounded on the concept of "low dose, low frequency of treatments, with consistency." You must not exceed recommended doses or frequencies of treatment. That one thing will keep you safe. I am not advising you to do this, I am just telling you that I do it, and I believe it can be done safely if you pay attention.

It is probably possible to "burn out" any gland. Then what? You would need to ask your doctor for replacement of the hormones you are missing, and that's a ride you don't want to go on. Please refer to Chapter 3: Safety. The other thing that will keep you safe is your attention to how you feel.

You *must* spend $30 and buy an automatic blood pressure cuff. You should know what your blood pressure should be, and what it is.

You should expect a "spike" in blood pressure and heart rate. How much would I say is acceptable? A nine-point increase in the top number would not cause harm for a day if your blood pressure were good, or well controlled, in the first place.

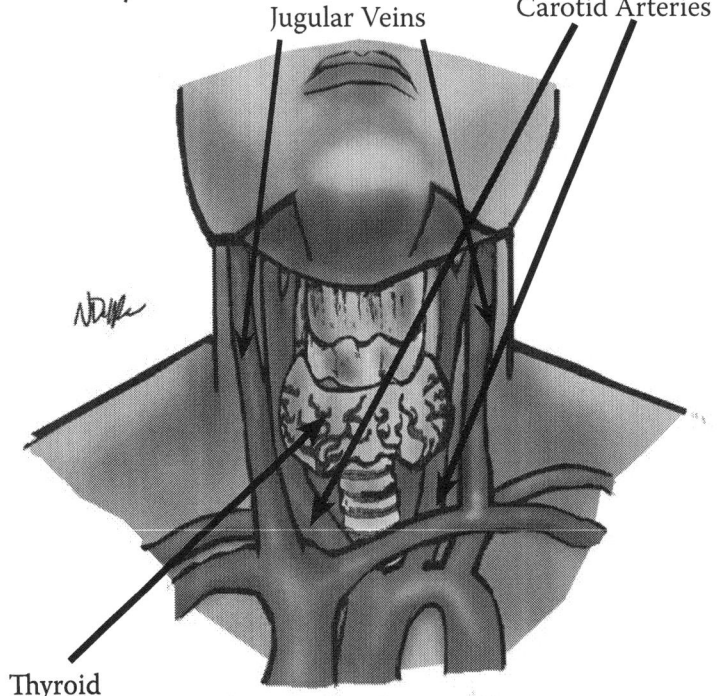

Figure 16: Blood Vessels of the Neck

This is very important! If your blood pressure goes up over ten points, and stays up for more than a few days after your initial treatment, then you must stop for two weeks.

If you want to try a lower dose, wait two weeks, and use ¼ the normal dose. As always, consult your physician, asking him/her to monitor your progress. This will not only protect you, it will help your doctor learn about this interesting technology. When he/she sees your improvement with no adverse effects, your doctor will take note.

When I speak of risks, only you can decide if you want to take the various risks I outline here. What are the risks? Well, if your heart is healthy, there should be virtually no risk. If you have a weak heart, then the sudden release of thyroid hormones could cause panic, heart arrhythmias, and in the extreme case, it may cause a stroke or heart attack.

If you are not sure about how healthy your heart is, then you should find a pulse higher up in the throat/neck, so that you are at the level of the jaw. Treating up there will avoid direct exposure of the thyroid gland; however, the thyroid will still be affected to some extent, so if you are not sure, start with ¼ of the recommended dosage time. If the results are not negative, then, a week later, try ½ the recommended dosage, and so on, once a week. Weekly treatments for any glands should stop after one month. If you have had no reaction, and you want to continue trying, then lengthen the treatment period to two weeks and slowly increase the dosage by 25% at a time.

I treat myself just below the voice box over the blood vessels to the right and the left of midline. I just *find the pulse and illuminate there.* That way, I treat not only the blood flow to the heart, but also the blood flow to the brain. I am also directly illuminating the thyroid itself to give myself more energy and to keep my weight down. Please refer to Chapter 16: Weight Loss. The carotid arteries carry blood up to the brain, and the jugular veins carry blood downwards to the heart. The next chapter covers the brain.

Be sure you are using iodized salt, because your thyroid's demand for iodine will probably go up. If that were to happen, and you did not have enough iodine, a goiter may possibly develop.

With a flat-tipped 5-milliwatt laser pointer, press into the skin lightly without hurting yourself, and hold it there for sixty seconds on the left, and sixty seconds on the right. If you are using a 100-milliwatt coherent light, illuminate the same two areas for ten seconds each.

Treat *only once per week* until you have reached your MMI (more endurance, more energy, less or no angina). Another signal to cut back is if you get a *blood pressure response*—a rise in your blood pressure of one to nine points in either the higher number (systolic) or lower number (diastolic). If it happens with even the first treatment, cut back to two weeks between treatments. After four weeks, cut back to two-week intervals between treatments, no matter what. Once you have reached your MMI, cut back by following the DelRe Protocols in Chapter 6.

Besides the risks I have covered, I have also come up with these hypotheticals:

1. Your thyroid gland may become dependent, and you may eventually need to take thyroid hormones. That is called *hypothyroidism*. (This is just hypothetical on my part; it has never been demonstrated or addressed in research.)

2. Your thyroid gland may become permanently "sped up." Once again, I am trying to come up with possible problems, and a person may possibly develop a hyperthyroid condition.

3. You may have some cancerous tissue there (I do not know the odds of that). If you do, the cancer may become stimulated and grow. Research implies that that won't happen, but who knows for sure? This has been specifically addressed in coherent light research, and they have *not been able to cause an increase in existing cancer cells in a living body*, but they have been able to do it in test tubes. Indeed, they found that in living bodies, the surrounding cells attacked and destroyed the tiniest of tumors via the immune response.[221] That sounds promising. More citations for related research can be found in Chapter 13: Female Health.

4. Be careful not to over treat by treating too often. It is theoretically possible to exhaust your thyroid gland. Once a week is maximum, and that should be cut back once you achieve a blood pressure response or positive response.

16

THE BRAIN[†]

Research has shown that coherent light therapy to the brain can have very positive results on stroke victims.[222],[223],[224],[225],[226],[227] One corporation is presently trying to get a device passed by the FDA to treat stroke victims within twenty-four hours of a stroke. It is a 100-milliwatt red coherent light.[228]

There have been studies that look at how much coherent light therapy power it takes to cause harm to the brain.[229] In those studies, they used coherent lights at powers of 7.5 milliwatts, 75 milliwatts, and 750 milliwatts/cm^2 over the skull of mature rats. In total joules (the unit used for actual power), they used 0.9, 9, 45, and *90 joules/ cm^2*, respectively.

Please remember, that, in normal circumstances, my maximum dose in this book is a mere 0.22 joules/ cm^2 unless for some rare reason, you are treating a smaller area for the same amount of time. In that case, it might a total of 1–2 Joules/ cm^2.

If this is confusing, do not worry. You really do not need to know it all. This information is for detail-oriented folks. I am trying to impress upon the reader the low dosages I am recommending in this book. In addition, your brain is over one thousand times bigger and heavier than a rat's brain.

The only level at which they could find any adverse neurological effects was at the level of 750 milliwatts/cm^2, and they were using dosages of up to 90 joules/cm^2! They were using a coherent light that went up to 5000 milliwatts. For part of the study, they had to increase the size of the laser tip from 2 mm to 3.5 mm because the tiny 2-mm opening caused the rats' skin to burn. *They wanted to see how much it took to cause a negative effect.*

This book uses a laser pointer of only 5 milliwatts for lay people and 100 milliwatts for professionals. I think you can see how much larger the dosages were in this study.

The researchers attribute the negative effects at 750 milliwatts as being caused by simple overheating, *not* by the attributes of coherent light itself. In other words, they tried to cause harm to the brain with a coherent light, but the only way to make that happen was to turn it up so high that it overheated the brain. Therapeutic dosages are not comparable to anything like this. Hold a laser pointer against your finger and see for yourself. Even a 100 milliwatt unfocused coherent light *will not* heat tissue, *even* if it is held in one spot, and I always recommend not leaving it on one spot anyway. You want to keep it moving, even in tight spots; a little movement during illumination is recommended. This book may seem reckless and liberal to the uninformed, but it is careful and conservative to the informed.

Coherent light therapy has been shown to help brain recovery from trauma.[230] Now, why would someone want to illuminate parts, or all, of their brain with coherent light? If I were recovering from a stroke, I would certainly use coherent light therapy over the area that is healing. If I were recovering from brain damage, I would use it.

If I were in the middle of a stroke, I would *not* use it, because the contraindications given in the NeuroThera studies listed hemorrhage (bleeding) as a reason not to do it. Seems like a good idea. Those studies also only included stroke victims reached within twenty-four hours. They received MRIs to rule out bleeding, tumors, etc. They want to market this device eventually and probably want it in emergency rooms. I would think their next step, after FDA approval, would be to get approval to treat stroke victims *after* they stabilize from the stroke. That is what I am talking about here.

Do not treat organs or glands if you have a fever. You could cause the fever to go up. With a fever, the body is already in a hyper-stimulated state. Do not add fuel to the fire.

DIRECT TREATMENT OF THE BRAIN†

(HIGH CONFIDENCE, LOW RISK)

With a 5-milliwatt laser pointer, illuminate each 3 x 3-cm area of concern for three minutes. That is 0.9 total joules, which equals 0.1 joules/cm². I suggest you limit it to three areas. Illuminate each 3 x 3 cm area for ten seconds with a 100-milliwatt coherent light. Limit initial treatments to once per week. *You have reached MMI* when you have noticed *any* difference. Decrease treatments per the DelRe Protocols in Chapter 6.

You can also illuminate tender spots, such as the *crown chakra*, a major part of Ayurvedic medicine in India. It corresponds in anatomy books with the *bregma*, the name given to the point at the top of the skull where the three big broad top skull bones all meet.

It is at the top center of your skull. Ideally, it should not be tender. If it is, I treat myself. Why? It is a sign of a "hot" acupuncture point. To me, it is a sign that something is stressed, and there is inflammation. I want those cells to start "marching" again and normalize.

With a 5-milliwatt laser pointer, treat a 2 x 2 cm area for two minutes. That will produce 0.067 total joules, which equals 0.168 joules/ cm² for this 2 x 2-cm area. The tender spot goes away. It has been months since I have done it, and checking now, it is still good. Treat the area for ten seconds with a 100-milliwatt coherent light.

INDIRECT TREATMENT OF THE HEART AND BRAIN†

(HIGH CONFIDENCE/HIGH RISK)

(Please refer to Chapter 15 for directions and protocols.)

Why would I want to treat the brain *indirectly* via the bloodstream? I really do not want to hit you with all the footnotes and references just given in the previous chapter on the heart and blood, so if you want proof, you can look there. Coherent light therapy has been shown to be very helpful to the health of the smooth lining of the blood vessels.

The reason I do it about once or twice a month is that I want good, healthy blood vessels in my brain. In addition, I want to carry the bio-stimulation effects to my brain to keep it young. I am also treating my heart at the same time for the same reasons. In addition, since I include the thyroid, I am keeping my weight down. I am not afraid of this because I am informed, and I have seen nothing negative in myself. The research shows nothing negative.

One danger I see is a person being so weak that they cannot physically handle the increase in thyroid hormones produced. Another danger is that they may have high blood pressure that is difficult to control, and this could drive it higher for a few days.

As stated before, the best way to avoid that is to treat as high up in the neck as you can find a pulse. You will avoid directly illuminating the thyroid gland, but you will still stimulate it somewhat via the indirect route and because it is so close. If it is a concern, start with ¼ dose, and slowly increase with each succeeding treatment.

17
WEIGHT LOSS†

TREATMENTS COVERED IN THIS CHAPTER ARE MODERATE CONFIDENCE/HIGH RISK.

My father was obese. He was 5' 9". He weighed in at over four hundred pounds for much of his life and was over three hundred pounds when he passed away. When I was in my early twenties, I saw a picture of him in *his* early twenties, and he was skinny like me. I knew I was in trouble when I took a night job with one of the first Domino's Pizza restaurants and started gaining weight. I have had to fight it since then. I have accepted that I cannot eat like other people. If I eat three square meals a day, I will gain weight. Therefore, I do not. I graze on mixed nuts through many days, because I have low blood sugar, and I do not want the ups and downs of that to be expressed. Nuts keep me sane.

Although I am not obese, I have been overweight for the last half of my life. With the coherent light, I would say that I have lost twenty pounds. My appetite has decreased. I have not treated myself on a regular basis. I presently illuminate the throat/thyroid about one to two times per month, just to see how it will work. I am still losing weight, bit by bit. I have about five more pounds to go. I will eventually go to four- to six-week intervals, hopefully stopping altogether.

If you are very weak and physically unstable, you may start with a five-second dose with a 5-milliwatt red laser pointer. It will do something, but it will be imperceptible, with no blood pressure reaction. You could safely do this once a week for two months, because five seconds of 5 milliwatts is a very tiny amount of light energy.

"Speeding up" the Thyroid[†]

Coherent light applied to the thyroid gland increases the production of thyroid hormones.[231, 232] Thyroid hormones speed up your metabolism.

I believe a person may be able to use coherent light therapy to "level the genetic playing field," as long as they are careful and follow the protocols in this book. I believe I have done just that. Do not look for a sudden loss in weight. It does not work that way. It will take two to four weeks to start seeing progress. Remember: persistence and consistency wins the day here.

If you are overeating because of self-destructive tendencies, this will not work because you may compensate by eating even more. This technique works, but you must control your eating. It does seem to help decrease appetite, but if your sub-conscious self-image wants you to stay overweight, you may overeat because you think you can get away with it. If eating too much is your habit, you need to learn to consume smaller portions, and your decreased appetite will do the rest. This is not magic. It needs your cooperation.

Blood Pressure Response[†]

You should *lengthen the period between treatments when you get a "blood pressure response"* (when your blood pressure goes up one to nine points for a day or two.) *The less of a reaction, the better.*

You should wait two weeks before the next treatment. If your blood pressure goes up *more than ten points* for more than a day or two, *cut back on the dosage to ½ the dose.* If you are prone to panic attacks, this will amplify any blood pressure reaction. After your first few sessions, you will know how you react and adjust for it. I am giving low dosages here, so you will be all right.

Do not illuminate the thyroid if you have a fever. You would likely cause a spike in the fever. With a fever, your body is already in a hyper-stimulated-state.

You will be treating the heart and brain indirectly while you are doing this. Please review Chapters 14 and 15 if you are going to use coherent light therapy to lose weight. You should read all the chapters in this book, because I have spread important information throughout this entire book, and you do not want to miss anything. (It is not a long book; you can do it!)

Please refer to Chapter 15 for the "Indirect Treatment of the Heart and Brain" section. You will be using the exact same procedure to treat yourself for the thyroid. Your instructions are there.

Buy and use an automatic blood pressure cuff. Be aware of any adverse side effects. You will most likely have a blood pressure "spike," and you must know what it is and how long it lasts. *If you take medication for high blood pressure, it must be under control when you start this.* Research has shown that coherent light therapy helps to lower blood pressure,[233,234,235] but that is certainly not the case when it is done through the thyroid. In time, since the blood supply to the brain and heart is being illuminated at the same time, you may see such blood pressure lowering effects. You may not, though, because of the effects of a stimulated thyroid gland. The "spike" in blood pressure may last an hour or two and may be followed by a mild to moderate increase the rest of the day.

Remember: A one- to nine-point increase in the top or bottom numbers would not cause harm for a day if your blood pressure were good, or controlled, in the first place. This is very important! If your blood pressure goes up over ten points on either the upper or the lower number and stays up for a few days, then you should wait two weeks and try with ¼ the recommended dosage time. Also, you just may not need this. Always consult with your physician.

In the beginning, this treatment should be considered high risk because everyone is different, and you do not know how sensitive you are. If you do any Internet searches on low-level laser therapy, and you look for contra-indications, you will always find the thyroid mentioned. The reason for this is covered in the previous two chapters, which you *must* read if you are going to do this. If you have a weak heart, or are low on electrolytes, the increased thyroid hormones in the blood may cause a sudden increase in heart rate, skipped beats, or blood pressure rise that could cause a heart attack or stroke. *If you were to ignore my guidance and treat it more than I recommend, you could theoretically burn out your thyroid gland.* Please refer to Chapter 15: Heart and Blood and Chapter 20: The Adrenal Glands for these and other details about positive and negative feedback loops.

Well, then, why in the world would I be telling people about this? Because if you do not have a very weak heart, if your blood pressure stays within a safe range, and you *stay within the guidelines here*, you can use this technique to lose weight and gain energy.

This is a homeopathic approach. Although it is not proven, I am expecting that the body will get used to the higher levels of hormone coming from a healthier thyroid gland and raise the threshold at which it stops sending the thyroid-stimulating hormone TSH, thus permanently raising the "thermostat" to a more normal level. We are going for the same effect that psychiatrists look for when they temporarily put someone on an SSRI antidepressant. They want to get the brain used to higher levels of serotonin, then hopefully take the patient off the medication, and see if the levels stay higher because the body is accustomed to higher, more normal levels now.

Not bad, if you are careful and attentive.

18

THE ABDOMEN[†]

It has been shown in studies that coherent light therapy to *stomach and duodenal ulcers* has...how did you guess? A positive effect.[236, 237, 238, 239, 240, 241, 242, 243]

STOMACH AND DUODENUM[†]

(MODERATE CONFIDENCE/MODERATE RISK)

Refer to figure 17. You are going to examine and illuminate the red area (17.A.) This area is also over the celiac plexus. The celiac plexus is the only place outside of the brain and spinal cord that has nerve cell bodies with nuclei. The Chinese call it the "second brain." The solar plexus chakra is here.

Make sure your physician has checked you out, and you know that nothing more than ulcers are there.

Lie on your back, knees bent, with both feet on the floor. Find the lower edge of your ribs. Follow them to the center where

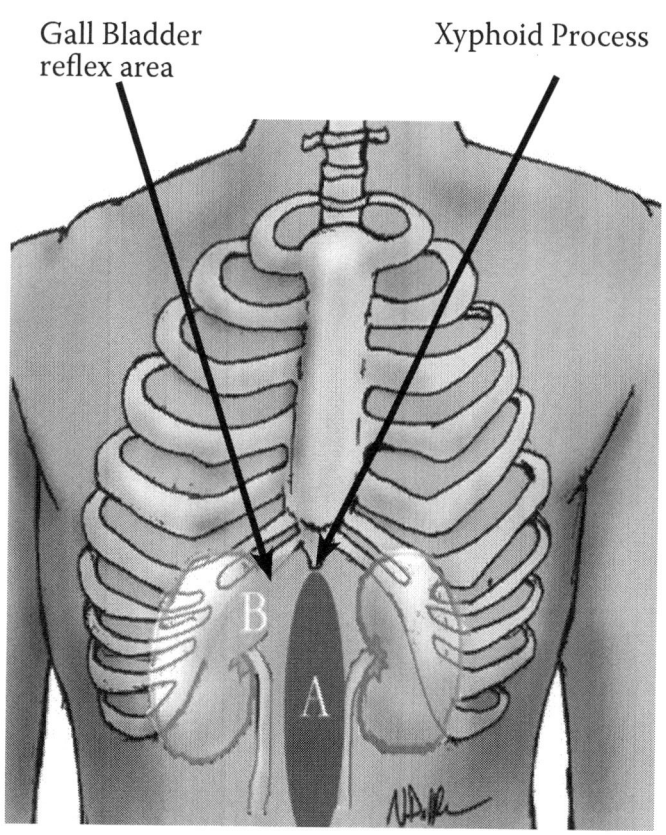

Gall Bladder reflex area

Xyphoid Process

Figure 17: Gallbladder and Stomach

your "wishbone" is. There should be a "pointed bony bump" there, pointing down towards your navel. That is called the xyphoid process. Go down to the tip and just past it into the softer area. You are looking for that spot. Press into that soft spot. Is it tender? How much? If it is tender, take note, and continue probing bit by bit, towards the navel. Press deeply without harming yourself. If your doctor has examined you, and you know you have a common stomach or duodenum problem, illuminate those tender areas with a flat-tipped 5-milliwatt laser pointer. You will want to illuminate each 3 x 3 cm area for four minutes. Illuminate for ten to twenty seconds per 3 x 3 cm area with a 100-milliwatt coherent light.

GALLBLADDER[†]

(MODERATE CONFIDENCE/LOW RISK)

You can find some information about your gallbladder[244] by pressing in, starting near (not on) the xyphoid process, and probing from there, up against and under the right bottom rib, probing along/under that lower rib to the right about 2–10 cm. Please refer to figure 17.B. If it is tender, illuminate the area for three minutes with a flat-tipped 5-milliwatt laser pointer, moving along under the rib, as deep as you can without hurting yourself. Be careful, too much pressure or probing could inflame your gall bladder. If this happens, it is because of the mechanical irritation applied, not the coherent light. If using a 100-milliwatt coherent light, illuminate the area for twenty seconds.

Check other areas of the abdomen; check the lower abdomen as well. Pay particular attention to your lower right area, deep into the front of the pelvis, to the right of the bladder. The ileocecal valve is here. It can become dysfunctional, and the coherent light can release it. This can help with constipation.

Illuminate any areas of tenderness. You have been checked by your doctor, and you know that no cancer or infection is going on there, haven't you?

Treat these areas starting at once per week (no longer than 4 weeks), and cut back to a two-week interval as soon as you see a response.

POSITIVE RESPONSE[†]

For illumination of glands or organs, consider *any positive response to be the point at which you decrease the treatment frequency.* If you get a negative response, you stop the treatment or try again two weeks later with ¼ to ½ the recommended dose, depending on how negative the response was.

When you have any positive response, go to the next lower frequency of treatments:

1. Start at once per week until positive response is reached. If it is not reached in four weeks, I still want you to go to #2 below.
2. Decrease to once every two weeks until MMI is reached. If it is reached the first time, go to #3 below, and so on...
3. Decrease to once every three weeks to reach, or stay, at MMI.
4. Decrease to once every four weeks to reach, or stay, at MMI.
5. Decrease to once every five weeks to reach, or stay, at MMI.
6. Decrease to once every six weeks to reach, or stay, at MMI.

Keep trying to decrease treatment frequency by increasing the interval between treatments. I think six weeks is a great goal. Try to see if you can last that long between treatments. Do it gradually, as laid out above. You may find out that you do not need to treat the condition any more.

19

TEETH, BONE, GUMS, TMJ SYNDROME[†]

When I started investigating coherent light therapy, I was amazed at the amount of research done overseas by dentists and dental surgeons. My first two years in the U.S. Air Force were spent working in the base dental clinic, so this is very interesting to me. (Military dentistry is very good. They have a lot of time to work on you, compared to civilian dentists.)

Coherent light therapy has been shown to:

- Decrease dental hypersensitivity.[†, 245, 246]
- Help with pain from orthodontic adjustments.[†, 247]
- Help teeth heal after drilling and filling.[†, 248]
- Have a positive effect on dental pulp.[†, 249]
- Help with pain after endodontic surgery.[†, 250]
- Help bone heal better and stronger after dental implant surgery.[†, 251,252,253,254,255, 256]

I used to assist in oral surgery while I was in the U.S. Air Force. I also assisted during root canal procedures. Although I have no wish to minimize the diagnostics involved in dentistry regarding root canals, what a dentist often ends up doing is to give a simple tap on the chewing surface of the suspect tooth. This quickly tells the dentist which tooth is "bad." The patient will jump when you tap the right tooth.

I have tried to find research to back up what I am about to say, but I could only find anecdotal stories. I have one—me.

I had a root canal procedure done on a tooth that had no damage or decay. The dentist told me that since it had been filled in the past, eventually, the pulp (nerve) can go bad anyway via amalgam or composite poisoning.

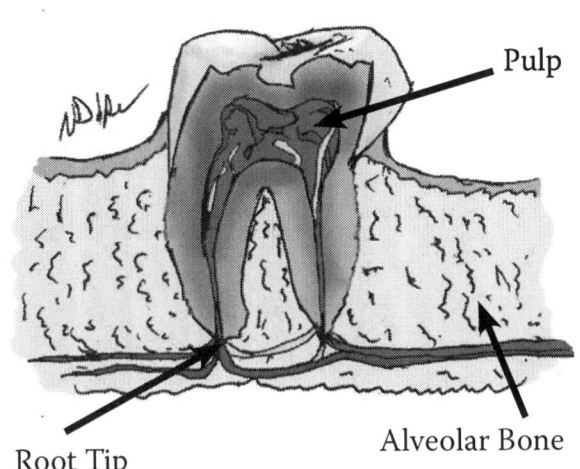

Figure 18: The Tooth

A year later, the same thing happened to the mirror tooth, but then I had my coherent light. I tapped it with a small metal rod I have, directly into the top/biting surface of the tooth towards the root tip. It was painful, the same as the tooth that needed the root canal a year earlier. (It could also mean a number of other things, but my dentist had recently examined me.) I illuminated it with my coherent light. The next day, the pain was gone. It gets a little sore sometimes. Guess what I do? There is no damage or decay, so other than an "old pulp," it is chugging along OK. Actually, that is wrong. It seems to be healing actually, because it has been a long time since I have had to treat it.

My mother, who has had many root canal procedures done, mentioned to me one day that she could tell she was going to need a root canal on another tooth. She illuminated it with her 5-milliwatt red laser pointer, and the pain was gone the next day. *Please realize this is not going to help with decay or infection.*

The pain from tapping the tooth may come from a possible abscess. I have seen quite a number of abscesses, each time a single one hanging from the end of a root tip when a dentist extracted an infected tooth. They usually range from 2 mm to 4 mm, looking like a tiny oblong pea filled with pus. Coherent light is not going to help something like that, and it is commonly present in someone who needs a root canal.

Antibiotics should not be delayed, so you should only try the coherent light therapy at the first sign of a problem, before any infection sets in. You can easily be too late. If you have had bad pain for a few days, or if you do not get relief in one day from the coherent light, get to the dentist. If you do get relief in one day, it was probably an inflamed pulp with very little infection yet. You should still see the dentist, as you may have a cracked filling that needs to be replaced, or something else may be wrong. I have seen plenty of pulps as well. They look like a sliver of raw oyster. They are primarily made up of nerves and capillaries with layers of fibroblasts, mesenchymal cells, and osteoblasts (bone-forming cells.) With a root canal, that pulp is pulled out by way of a barbed "needle," and replaced with a type of rubber, effectively killing the tooth. Even if I needed antibiotics, I would still try a course of coherent light therapy before deciding to "kill" the tooth with a root canal. It has recently been discovered by the Japanese that dental pulp is a rich source of stem cells. The teeth have plenty of resources to heal.

BONE†

The type of bone that holds your teeth in is called alveolar bone. It is "spongy" bone. It has microscopic pores, holes, like a sponge. If you are experiencing bone loss, you should be illuminating it with coherent light. You should also stop smoking and start taking Vitamin C with bioflavonoids or green tea. You can see the references on this chapter's first page regarding tooth implants. If you have had a tooth implant procedure done, you should certainly be illuminating the bone that was implanted. Not only do we have specific implant research on the previous page to back us up, there is *even more* positive research regarding coherent light therapy and *bone repair* in general.[257, 258, 259, 260, 261, 262, 263, 264, 265, 266]

FOR GENERAL HEALTH OF ALL TEETH, BONE, AND GUMS:

First, let me say that if you have gingivitis (gum problems), it is usually due to smoking, poor cleaning/flossing, bone loss, or Vitamin C/bioflavonoid deficiency. When I worked in an Air Force dental clinic, we repeatedly saw a bulimic female dependent with severe gingivitis and cavities. The word and condition *bulimia* was not common then. You need to attend to any of the above issues, if applicable. Applying coherent light to the teeth, bone, and root tips will naturally treat the gums. I treat all my teeth a few times per year, just because I want strong bone, healthy pulps, and good circulation there. I treat any *troublemakers*, if, and when they appear.

I was nineteen years old when the Air Force sat me down with a dental assisting home study course to assist a brand-spanking-new dentist/captain named Glenn Chester Prezkop. (I included his middle name because he has often remarked how much he likes it.) Anyway, one thing he emphasized was that when it comes to your teeth, nothing is better than keeping the original thing. That still makes a ton of sense.

HOW TO ILLUMINATE THE TEETH†

(HIGH CONFIDENCE/LOW RISK)

Please refer to figure 19. See the red areas above and below the teeth? This is where you want to illuminate. The roots are generally two times longer than the exposed part of the tooth. *Your main goal is to hit the root tips and surrounding bone.* That is where the action is. You will inadvertently treat the teeth, bone, and gums when you go after the root tips. I call it *collateral repair.* The same phenom-

87

enon has been seen with laser surgery. The surrounding tissues heal faster.

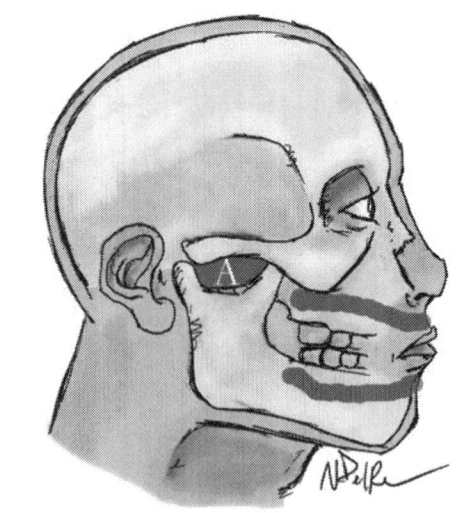

Figure 19: Illumination of the Teeth

Cover the front half of your 5-milliwatt laser pointer with thin plastic wrap to keep it clean and dry. You will do this in front of a mirror. *Be aware of eye hazards,* including *reflections* from the *mirror! This is the first time I have spoken of any treatment around the face, so be careful not to look into the laser pointer.*

Start with the upper molars on the left or right. Without injuring yourself, position the tip of the flat-tipped laser pointer so that it is up underneath your lip and cheek, up as high behind your deepest molar as you can go, relatively *flat* against the gum/bone. You want to be coming in directly/horizontally from the side, if possible. If, for the back molars, it is too difficult to lift the lip/cheek enough for the laser pointer to be horizontal, just use a slight upward angle. *Do not hurt yourself.*

Illuminate each spot for ten seconds, and move 1 cm at a time, stopping at each spot for ten seconds. Repeat this until you illuminate all of the upper teeth's root tips. Do the same for the lower teeth. If it is too difficult to do from the outside, you can illuminate your teeth from the inside of your mouth as well. The maximum total dosage time should be less than two minutes per row of teeth.

For a 100-milliwatt coherent light, use two seconds per spot, and the total time per row of teeth should be less than thirty seconds. You can go through the cheek.

If you have specific problems you are trying to overcome, the frequency of treatment should start at once per week for a month, and then cut back to one treatment every two weeks. Your positive response should have leveled off at this point, and you should cut back further per the DelRe Protocols in Chapter 6.

If you are sensitive to the coherent light, you may notice some skin irritation. Cut back on the dosage by half next time. You may find that any future sensitivity will go down as your body becomes accustomed to the coherent light.

Root Canal Avoidance[†]

(Moderate Confidence/Low Risk)

Hold the flat-tipped 5-milliwatt laser pointer just as described. Position it at the root tip of the offending tooth. Illuminate that area, from the root tip to the crown of the same tooth, scanning slowly for sixty seconds. Spend most of your time at the root tip. You can also illuminate from inside your mouth, through the upper bone/palate, or lower mandible. Do this from inside the mouth for thirty seconds. Even though you do not have much space, keep the laser pointer moving in small segments.

If using a 100–milliwatt coherent light, illuminate for fifteen seconds. You can go through the cheek, and you should not need to treat from the inside of the mouth because of the increased penetration from the 100-milliwatt light.

If you are sensitive, you may notice some redness/irritation in some areas in the face for a few hours to a day. If so, cut all dosages in half next time.

If you are at MMI after just one treatment, you will not need to follow the DelRe protocols. If it gets sore a few days or weeks later, and you have been to the dentist, treat it *once per week*, and then follow up per the DelRe Protocols in Chapter 6. *Do not ignore pain. There is a high likelihood of infection*, and that is never something to ignore. If the tooth does not feel significantly better the next day after the first treatment, get to the dentist quickly. Once you have antibiotics, you may still be able to use coherent light to save the tooth on your own if it has no cavities or cracked fillings. Wait for the antibiotics to clear up the pain, and then start with the coherent light therapy while still on the antibiotics.

DENTAL IMPLANT SUCCESS[†]

(HIGH CONFIDENCE/LOW RISK)

Hold the flat-tipped 5-milliwatt laser pointer just as described in the root canal avoidance procedure above. *Position it at the "screw tip" of the implanted root (this should be about the same depth as the original root tip).* Illuminate that area, from the screw tip to the gum line of the same tooth, scanning slowly for sixty seconds. Spend all of your time at the gum line, implant, and its deep tip. You can also illuminate from inside your mouth, through the upper bone/palate, or lower mandible. Do this for sixty seconds. Even though you do not have much space, keep the laser pointer moving in small segments. Treat twice per week for four weeks, cutting back to once per week for the next four weeks. The bone should be mostly healed by then. You may want to illuminate the area periodically to keep it strong for the future.

If the bone is eventually stronger and thicker there, it may change your occlusion (bite). If you notice anything like that, see your dentist so he can sculpt the biting surface of a few teeth.

For a 100-milliwatt coherent light, use a treatment time of fifteen seconds; you should not need to do the inside, and you can illuminate right through the cheek.

Temporomandibular Joint Syndrome, ("TMJ" Syndrome)†

(Moderate Confidence/Low Risk)

TMJ syndrome is in the eye of the beholder. *The dentist sees* malocclusion and nocturnal bruxation as the cause (teeth that do not mesh well when biting, along with grinding your teeth at night). As a result, they often sculpt down suspect cusps, give muscle relaxers, and fit their patients with nighttime bite guards. The worst-case scenario is a dental surgeon that suggests cutting a section of the mandible (jawbone) out.

The psychotherapist sees a person grinding their teeth at night because of stress. They try to treat the mental causes of the stress and probably send many patients to dentists to make nighttime bite guards for relief in the meantime.

I have not seen much real success with bite guards, unless the person just cannot be stopped from grinding their teeth at night.

The chiropractor sees misalignment of the upper cervical spine, including the vertebra that joins the skull with the neck, as a major cause of TMJ syndrome. I used to have TMJ syndrome, and it had been very bad for as much as a year at a time. I have not had it in over fifteen years. (It would have been interesting to try the coherent light on it, though! More on that after the following information.)

How did mine go away? Well, when I was in the Air Force, five days of taking a muscle relaxer took it away. Then, many years later, in my first years of being a chiropractor, it came back, big time. It would seem to fade, and then I would simply be eating and get that big "CRUNCH" in one of my jaw joints like a rock that had shattered in my jaw joint, and it felt that painful. There is something like a disk there, and it was probably part of the crunch. Then I would think, "Oh, here we go again." I would have pain with all eating for the next month or so, until the next cycle. I would have to eat very gingerly. I believe that in my case, it had to do with a misalignment of my upper cervical spine, and the right chiropractor finally worked on my spine, and it went away for good. I have personally helped many people with this problem. Typically, the first cervical vertebra has rotated such that its right or left "wing tip" is underneath the TMJ on the most painful side.

Regarding my patients with TMJ syndrome, many of them have already gone the dental route. If dentistry fixed their TMJ syndrome, they of course do not report a problem to me. I would imagine it is the same way with dentists.

If a patient is grinding his/her teeth at night, I give him/her a calcium/magnesium supplement in the normal 2:1 ratio found in most bottles. The teeth grinding usually stops the first night. These people usually have restless leg syndrome as well.

I must mention *restless leg syndrome* here. Taking a restless leg syndrome *drug* is like putting black tape over a trouble indicator light in your car that says "Hot." Your car is low on water, not black tape. In other words, your body is low on calcium and magnesium, not a synthetic chemical drug. The magnesium is a natural muscle relaxer, and single atoms of calcium are what provide millions of sub-microscopic "doorstops" in your muscle cells that stop the natural tendency of muscles to contract. In order to contract, a nerve impulse ends up knocking those millions of doorstops loose, and the doors automatically close—your muscle automatically contracts.

If you are low on calcium/magnesium, your signs may be one or more of these: *heart arrhythmias (palpitations, skipping beats) muscle cramps, muscle twitching, restless legs at night, jerking just as you fall to sleep, or grinding your teeth at night.*

If a couple of these things are happening, your body is slowly robbing the calcium bank in your bones. Once you are over twenty years old, it becomes exceedingly more difficult to make new deposits to that bone bank as the years go by. Occasionally a person needs potassium for *muscle cramps alone.* Too much potassium by itself over time will likely cause a deficit in calcium/magnesium and actually cause the problems listed above, particularly muscle cramps. Contrary to conventional medical wisdom, you do not want to use potassium for leg cramps, unless you are one of the 5% that calcium/magnesium does not help.

In 1983, I did a research paper in chiropractic school called "The Nutritional Treatment of Geriatric Patients with Cardiovascular Disease." In the research journals and books, *going back one hundred years,* I found the same thing repeatedly—*calcium and magnesium prevent heart attacks and strokes in a big way.*

They repeatedly gave similar ratios—2:1—not so much different as found in nature. In other words, for every 500 mg of calcium, your pill should also have about 250 mg of magnesium. For different size pills, it is easy to figure out. Your body uses a huge amount of calcium per day, more than any other elemental nutrient—1000 milligrams per day on average. Calcium must be taken with magnesium, and vice versa. This is because they work together, and an increase in one will cause the body to demand more of the other. Maybe some day "Brand Name" calcium companies will figure that out, invest a few pennies of magnesium in *all* their overpriced bottles, and start really helping the people that have fed them millions of dollars over the years, instead of potentially throwing their customers' calcium/magnesium out of balance, possibly inviting heart arrhythmias and other problems.

One other thing, it is well known that water softeners remove minerals from water. Water softeners have been repeatedly indicated to be a cause of sudden heart attacks, due to lack of calcium and magnesium. (They usually blame sodium.) One researcher concluded, "Health education could be used to discourage the use of water softeners to treat water used for drinking and cooking."[267] What about all the other minerals? Zinc, for growing young men, chromium for the young pancreas trying to keep up with all the sugar, calcium for bones, teeth, and muscles, and all the rest—gone with the water softener. Who cares out there?

Back to TMJ syndrome. Considering all of the above, one component of TMJ syndrome was not mentioned: the pterygoid muscles. There are two pterygoids involved with each jaw joint, and they are difficult to get at. In my opinion, the relaxation of the pterygoid muscles is a big key to muscle relaxer success. The pterygoids can develop trigger points inside their muscle bellies and become spastic. One way to stretch them is to bite on your finger(s) with your fingertip back behind your last molars. Therefore, instead of biting your fingertip, you will be "gumming" your finger tip. When the jaw closes, the pterygoids are exposed, and you know it by the pain they produce when stretched/challenged by your fingertips. When you are in this position, push your fingertip up and back into the soft, tender tissue. Stretch it as much as you can stand without injuring yourself. When you are done, you should notice immediate relief, maybe for good, maybe for just a while.

You can also reach the pterygoids with a coherent light. (Geez, I thought I would never get to this.) Refer to "A" in figure 19.

With a 5-milliwatt laser pointer, treat that area for two minutes. Keep it moving in that area. You can also treat from inside the mouth in the area where the fingertip was. Point the laser pointer up and back. Do this once a week for 2–4 weeks, depending on results. If it does not help, then stop. If it does help, then you should stop unless it comes back. Treat that same area for ten seconds for a 100-milliwatt coherent light

Although there has been some positive coherent light research into tinnitus (ringing of the ears), I have seen enough negative studies that I think you should stay away from the ears. It probably affects the little bones of the inner ear. If you notice any ringing of the ears, you should stop any further self-treatment.

20

THE ADRENAL GLANDS†

If you have high blood pressure that is difficult to control, or water retention issues, ***do not use this chapter.*** *It* **will** *drive your blood pressure up.*

As I have mentioned before, taken together, the crossing hormonal systems of the body are like a *"liquid nervous system."* Both as an electronics technician in the U.S. Air Force and as a doctor, I have rarely seen any textbook cases. That is because so many little things affect other little things, and they can sometimes all cumulatively end up in the technician's/doctor's confusion. So it is with the "liquid nervous system." There are many *positive and negative feedback loops* (checks and balances) in your nervous system and your "liquid nervous system." It is like an electronic device. (Alternatively, is the electronic device imitating nature?) So many elements affect so many others, that at a certain point, you cannot really trace it all, although the researchers try, and we do learn more all the time. We get more data, which leads to more complexity and more confusion...and more answers. The hormonal glandular system that injects hormones directly into the blood is called the *endocrine system.*

If you have noticed, a many doctors who specialize as *endocrinologists* here in the U.S. are from India. This is likely because the glands correspond to spiritual centers called *chakras* in their religions and traditions, going back thousands of years. I believe Indian endocrinologists would be intuitive in glandular balancing.

Endocrinologists may be shocked by what I am telling you in this book, until they read it and try it themselves. A lack of knowledge can be frightening.

This is my argument: Nobody knows everything they are doing, because we just cannot accurately predict all the actions and interactions and interdependencies of all the hormones of the body. The endocrinologist makes his/her best educated guess, and if

93

that does not work well, he/she adjusts it until the best level of success is achieved. What will happen today, or tomorrow, or next week, or next year with any hormonal therapy? Nobody knows for sure, because every patient is different. The endocrinologist's beginnings are based in science, but his/her main allies are time and experience.

If you increase the level of *cortisol* in the blood, it has its own negative feedback mechanisms which "down-regulate" its production. ACTH (adrenocortotropic hormone) influences the adrenal glands. ACTH stimulates the release of *cortisol*, described below. Here is how it works: The **brain** "sees" low cortisol, (a need for more cortisol,) in the blood, at a particular threshold. It turns on a "liquid switch" called *ACTH, which tells the adrenal glands to produce more cortisol.* When there is "enough" cortisol in the blood stream, the brain turns the switch off. So here is the rub: if you increase the production of cortisol by way of coherent light therapy, will the brain and its "liquid switch" not decrease it? Yes, to a level. Then, why talk about this? Because we are talking about making weakened adrenal glands healthier, that is all. We are not trying to manipulate them. Make them healthier, and let the technology of God do the hormonal balancing. He is good at it.

To be clear, the treatments in this chapter will likely affect your system in some way for one to three days, with higher blood pressure being the immediate concern. Compared to illuminating the thyroid, which also has a blood pressure reaction, *the adrenals have a higher blood pressure reaction, and it lasts longer. So stay within the guidelines, monitor and record your blood pressure, and adjust the dose accordingly.*

Only one treatment can rejuvenate your adrenal glands so effectively that *they react even more with the next treatment,* so always monitor your blood pressure, even if you think you know what to expect. *Spread out the treatments as soon as possible to avoid adrenal exhaustion.* Although this therapy is used to help adrenal exhaustion, *it can also cause it* if you ignore my guidelines and treat them too often. *If you are prone to panic attacks,* that will contribute to the blood pressure reaction, so you should start with a ¼ to ½ dose.

Why do this? Because we are identifying stressed-out glands and attempting to rejuvenate them. If successful, they may not need to be rejuvenated again for a long time, maybe never. Always try to use as little stimulus as possible with fewer and fewer treatments over time. This is a homeopathic approach.

ADRENALINE[†]

The adrenal glands produce a number of different hormones. *Adrenaline* (epinephrine) is the most well known, and it comes from the inner part of the adrenal gland (*the medulla*). Adrenaline is the "fight or flight" hormone meant for danger response. It does much more. Your blood stream gets a shot of it every time you stand up.

This is because adrenaline constricts the blood vessels in your arms and legs when you stand up, thus squeezing it up into the brain, against gravity. Why is this important? Well, if it does not give that shot to the blood stream, you will want to faint when you stand up. This is because gravity pulls the blood down out of your brain and into the "over-relaxed" blood vessels in your legs when you do not have adrenaline-constricted blood vessels down there. That is a simplified version of it, but you get the point. It is called *postural hypotension.* If you repeatedly feel light-headed when you stand up, your adrenal glands are most likely "exhausted."

This is common today. The "fight or flight" function of your adrenal glands is meant for occasional use in the animal kingdom. In today's world, the "urban jungle" stimulates your adrenals hundreds of times per day by way of traffic, work, finances, the news, the demands of "being to the kids' soccer game on time," and the thousands of advertisements seen/heard daily that are meant to hit you in a deep, motivating way. Add coffee, nicotine, and excess alcohol, and you are beating up on those adrenal glands. Add a death in the family, a divorce, a job loss. Add the whites: sugar, flour, and salt, and, well, you know.

CORTISOL[†]

The outer portion of the adrenal glands (*the cortex*) produces *cortisol. Hydro-cortisone* is the synthetic form of cortisol. *Cortisol* does many things, and I must say that my information here is a gross simplification of the myriad interactions and bodily processes of cortisol (or any of the hormones mentioned in this book.)

1. *When your body and adrenal glands are normal,* cortisol helps to maintain blood sugar and blood pressure. It boosts the immune system, increases appetite, and decreases inflammation.
2. *When your body is stressed,* the adrenal glands produce *more cortisol.* In higher blood concentrations over time, it can cause:

- higher blood pressure
- water retention
- immune system suppression
- osteoporosis
- fat deposition around the mid-section of the body
- blood sugar problems
- increased appetite
- weight gain
- water retention

...as well as a list of other problems. Cortisol breaks down into a few things, including cortisone. Cortisone increases insulin resistance, and thus can cause high blood sugar (*hyper*glycemia) and diabetes. When your body is stressed too much over time, the adrenal cortex can also produce more *androgens*.

3. *When the adrenal glands become exhausted*, they produce *lower amounts of cortisol*, which can lead to low blood sugar (hypoglycemia), a weakened immune system, fatigue, low blood pressure, increased inflammation (aches and pains), as well as other problems, with the extreme case being Addison's disease, which is more than just simple adrenal exhaustion.

ANDROGENS[†]

Androgens are precursors to *testosterone, and are synergistic with it.* Androgens are also precursors, (building blocks,) for estrogen. In women, the adrenal glands are the only source of the male hormone testosterone. After menopause, the adrenal glands become the primary source of the female hormones estrogen and progesterone. If the adrenals are over-reacting with increased output in general, women can also experience an increase in male characteristics. The "rage" component of menopause could be explained by androgens. In a way, they are going through what young men go through when getting used to handling testosterone.

Back to adrenal exhaustion. This is all difficult to track and spot because a person with this type of problem probably fluctuates between one state/stage and the next for months or years before the adrenals decidedly break down to a worse level. An extreme sign would be falling into postural hypotension such that a person must crawl, otherwise they may pass out from standing up and walking. If you are in a condition such as that and getting good medical care, you should obtain the liquid colloidal minerals I mention later in this chapter. I have only seen one case that bad, and he was being treated for arthritis. With minerals, he was up and walking in two days.

My purpose here is not to manipulate the adrenal glands for any one specific hormone. My intent is to help you identify weakened adrenal glands, and to rejuvenate them with coherent light therapy before they get worse. They may go through a surprising burst in hormones, both the ones you want *and* the ones you do not.

Let me give an example: me. I illuminated my adrenal glands once per week for two weeks. The second treatment drew a larger and longer response. My blood pressure went up about twenty points on the systolic (top number) and about ten points on the diastolic (bottom number). It did this for two days. After my second treatment, the effects were stronger. My blood pressure went up thirty points on the systolic measurement (the top number) and lasted a week. I sure did have energy; I felt "pumped up." I also felt aggravated, and my blood pressure was too high. If I were someone who was "on the edge" of a stroke, it might have taken me over that edge. I took extra blood pressure medication on those days (I do not recommend that, and I do not take much in the first place). The doses given below are half of what I used on myself. If you are able to tolerate these doses well and are not seeing a clear blood pressure reaction in the one- to nine-point range after three treatment sessions, then you can increase the dose by 25% at a time until you see a response.

The proper response should be a <u>slight</u> rise in blood pressure (one to nine points.) The less, the safer. I call this the *"blood pressure response."* In addition, because the glands are getting healthier, you should see more of a response each time at the *same* dose. So, give it a few treatments before you decide to raise the dose. Do this once a week until you see a response in either how you feel or in the daily blood pressure measurements you have been recording on a calendar.

When you get a response, do not treat again for two weeks. You do not want to burn out your adrenal glands. *This treatment will likely cause temporary weight gain/water retention because of the cortisol.*

- Since cortisol and adrenaline raise blood pressure, you must own and use an automatic blood pressure cuff.
- You should expect an increase in blood pressure for one to three days.
- If it goes over ten points higher than normal, wait two weeks before you treat the adrenal glands again. Decrease the dose to 25%–50%.
- As always, consult with your doctor. Start with the low dosages given.

If you are very weak and physically unstable, you may start with a five-second dose on each gland with a *5-milliwatt red laser pointer*. It will do something, but it will be imperceptible, with no blood pressure reaction. You could safely do this once a week for two months, because five seconds of five milliwatts is a very tiny amount. I hope that you will notice a subtle increase in energy and stamina after a month or two.

If you were overweight, I would not recommend doing this, unless you have low blood pressure. Even so, you should stop treatment as soon as you get one blood pressure response. You will temporarily gain weight from this.

When you get a blood pressure response, no matter what, you will have fluctuations in your hormonal levels of not only the hormones I talk about here, but others as well. This can temporarily affect your system and moods in your own individual ways over the next few days following the treatment. Only you can decide if the effects are mild or not. *You must decrease the dose so that any side effects are mild and of short duration.*

Why would anyone want to rejuvenate the adrenal glands? Well, here are a few reasons:

1. POSTURAL HYPOTENSION[†]

(HIGH CONFIDENCE/HIGH RISK)

This tells you that your adrenal glands are weakened. When you are chronically stressed, one of the things that can happen is that your adrenal glands may enlarge in order to try to produce more hormones. It shows up in the later stages of chronic adrenal stress. That is bad. For fifteen years, I have occasionally given patients liquid minerals to support the adrenal glands for postural hypotension. It usually works in a day or two. It may get you back up but do not expect that your adrenals can heal that fast. If you decide to buy some locally, and they taste awful, then those are the old-fashioned kind that have been sold since the 1930's. They work if you can get past the taste. They are mined in Utah. The original good tasting brand is New Vision, which you can order by calling 1-800-MINERALS for $20 per quart bottle. They can also be found at www.newvision.com. I have also recently found an interesting new liquid mineral product at www.vitalearth.net. The product is called "Fulvic Mineral Complex". It has the added benefit of detoxifying to body of heavy metals, and pollutants. Fulvic is a molecule that works like EDTA. Fulvic is found at the roots of plants. EDTA has been used for many years in cases of heavy metal poisoning. It binds to heavy metals in the body, and they are expelled from the body. That is known for sure. In addition, many doctors around the world insist that EDTA helps clear the arteries of plaque, as well as clearing the body of heavy metals, pesticides, and pollutants. I tried it once, and it wiped me out so much that I stopped. Fulvic works the same way, but is not as harsh. The Fulvic Mineral Complex dropped my blood pressure for a while. It may have been a "detox" reaction due to the oral chelation effect. You should monitor your blood pressure if taking these, and cut back on the dose. I can only take ¼ dose per day, otherwise I feel exhausted. That is not typical of liquid minerals, but it is typical of oral chelation therapy. Start with ½ a dose.

The use of any of the liquid minerals above will likely increase your body's demand for calcium and magnesium. This is because there are millions of "little chemical machines" in your body, just waiting for those minerals. When the "machines" get them, it is like adding a needed gear that was missing. Then they really start working, and you feel the difference. When they start working, they need more calcium and magnesium. Take the minerals in the morning, and take your calcium/magnesium at night. One last thing, I am new to the Fulvic Mineral Complex brand of liquid minerals, but they are definitely different from the ones I have used for 18 years. The Fulvic Mineral Complex dropped my blood pressure initially. You should monitor your blood pressure if taking these!

You may feel "yucky" for a week initially, because fulvic does detoxify you at first. I have found that I can only take ¼ dose a day, and I don't take it every day, because I mainly use it for a light "salting" of my body with trace minerals. Actually, you may be better off just getting the regular liquid minerals. You won't get any surprises.

Traditional nutritionists have known for generations that raw adrenal gland will help support the adrenal glands. You can get this in any nutrition store in pill form, and it works.

You can also illuminate the adrenal glands.[268]

2. Chronic General Inflammation[†]

(Moderate Confidence/High Risk)

This section is for chronic inflammation "all over", or jumping from place to place in your body.

Cortisol and its brothers are your body's natural cortisones. They decrease inflammation in your joints and other tissues. When the adrenal glands are exhausted, they not only produce less adrenaline, they also produce too little cortisol in the stage of adrenal exhaustion. I believe that coherent light therapy can help to get those glands back on track.[269]

3. Post-Partum Depression[†]

(Moderate Confidence/High Risk)

99

It has been hypothesized that when a mother is stressed/depressed too much during her pregnancy, her adrenal glands are stressed/weakened, and in the last trimester, her body may be literally using the hormones from her unborn baby's adrenal glands. These babies are then born with over-reactive adrenal glands and are more likely to have behavioral/drug abuse problems later in life, like a "crack baby," or one born to a mother who smokes tobacco while pregnant (similar in most ways to a crack baby but less severe).[270, 271, 272] In the meantime, the mother cannot use the baby's adrenal glands after the baby leaves her body. This leaves mom in a depressed state until her adrenals rebuild themselves. Until they regain health, they do not respond to her ACTH "liquid switch" very well,[273] and the baby starts life in a bad way.[274]

There is evidence that the mother also has had this problem from childhood. She may have started life herself the same way, especially if her mom smoked during pregnancy.

Although you must not treat yourself during pregnancy, you can treat yourself afterwards for the baby blues. Alternatively, if you recognize these signs before you are pregnant, why not help your adrenal glands now?

4. CHRONIC FATIGUE SYNDROME (CFS)[†]

(LOW CONFIDENCE/HIGH RISK)

This likely will not cure CFS. I will not even try to go deeply into CFS in this book, other than to make some general comments. One theory that holds weight in my book is that some CFS is associated with Epstein Barr Virus/mononucleosis. If part of your symptoms are that you always feel like you are on the edge of getting the flu or a cold, then you are a candidate for what I am about to say.

There is a supplement known as dimethylglycine, (DMG.) The only form I have experience with is the sub-lingual form that comes in foil-sealed packets. The others may work. You can usually find it in health food stores or online. The same thing has been available under a variety of brand names for twenty years. You dissolve it under your tongue. You want to take three per day (spread throughout the day) for a few days until you feel the virus leave. Then cut back to one per day for a few weeks.

Although you may want to illuminate the adrenals to rejuvenate them, do not do this until you have been taking only one pill a day for at least a week because of possible allergic reactions.

Is it possible that one pill of DMG per day may exaggerate any side effects of coherent light therapy? I do not know but watch for any sensitivity. DMG is a great immune system stimulant, and it works particularly well against the Epstein Barr Virus as well as other herpes viruses. If you have the above symptoms, find some DMG, and take it as outlined above. You may notice that what comes out of your nose is very white. That is your immune system at work.

5. Asthma[†][275, 276]

(Moderate Confidence/High Risk)

With asthma, you _do not illuminate the lungs_ or the bronchi. You should probably stay away from the direct treatments to the heart. You may even find that the indirect treatments to the heart may cause an asthmatic flare-up. (Some positive asthma research does treat the bronchi,[277] but I think it is too much risk for home treatment.)

You, however, can indirectly treat your asthma by illuminating your adrenal glands. Will it work? I cannot say for sure in your case. Are there risks? Yes, most likely. Are they controllable? Yes, if you follow the conservative instructions below.

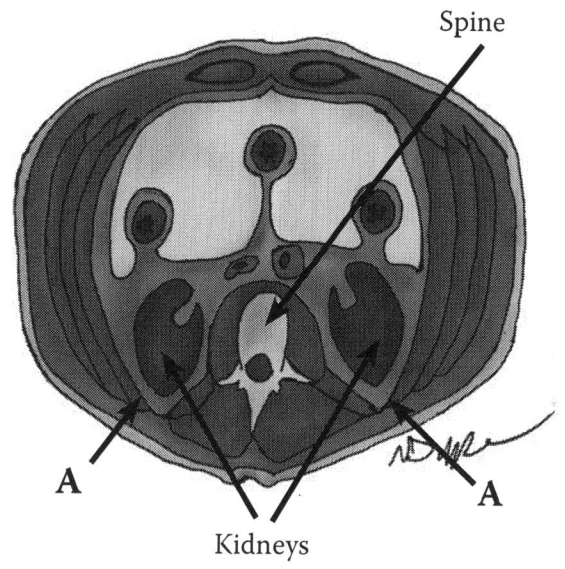

Figure 20: Abdominal Cross-Section

How to Illuminate the Adrenal Glands[†]

Refer to figure 20. See how close the kidneys are to the surface of your back. Observe their proximity to the spine. Look at the arrows labeled "A." Those indicate the direction and placement of the coherent light.

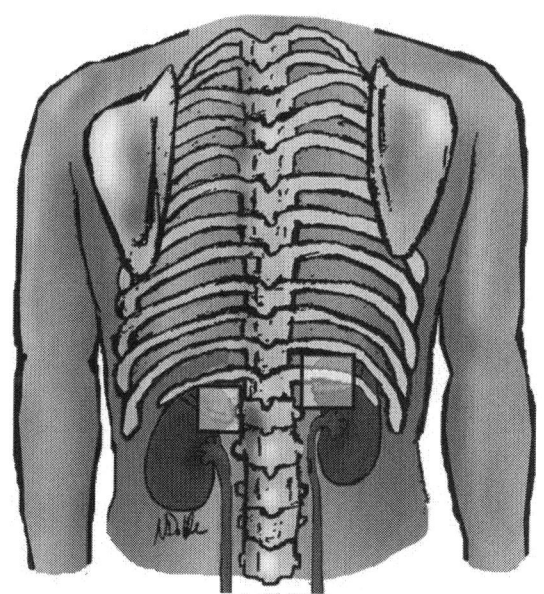

Figure 21: Adrenal Glands, Posterior View

Refer to figure 21. You are looking at the back. The adrenal glands sit on top of the kidneys. See how the right kidney is higher than the left kidney. Look at the squares drawn. You want to illuminate those areas. On the left, the square is just below the twelfth rib. You want to be as high as possible, yet stay below the lowest rib on the left.

On the right, you want to be 4 cm, (1½ inches) higher than the spot you illuminated on the left. (You will il-

101

luminate a spot between the ribs.) You must probe and find the lowest rib on yourself. You may need an assistant for this, as it is awkward to treat yourself. *If you are on blood pressure medication, start with one-half the dose given below.* ***If your blood pressure is difficult to control, do not illuminate the adrenals glands—do not use this chapter!***

For illumination of the adrenal glands, consider any *one- to nine-point "blood pressure response" to be your signal to decrease frequency of treatments:*

1. Start at one treatment per week, until you get a blood pressure response. If two weeks/treatments go by with no response, increase the dose by 25%. Once you get a response, go to #2 below. If you go four weeks without a blood pressure response, I still want you to cut back to #2 below.

2. Wait two weeks. Treat the adrenals with the same dose that gave you a blood pressure response at one week. If you have waited two weeks, treated again, and the treatment works, go to #3 below. Otherwise, go back to step #1 if you now are getting *no* blood pressure response. Once you get another blood pressure response, stretch it to two weeks again.

3. Wait three weeks. Treat the adrenal glands with your last successful dose. If you get a blood pressure response, go to #4 below.

4. Wait four weeks. Treat the adrenal glands with your last successful dose. If you get a blood pressure response, go to #5 below.

5. Wait five weeks. Treat the adrenal glands with your last successful dose. If you get a blood pressure response, go to #6 below.

6. Wait six weeks. Treat the adrenal glands with your last successful dose. If you get a blood pressure response, and you have achieved your goals, you can either stop or treat every six weeks if still needed.

Treat the *left* side first. *You want to aim for the squares in Figure 21.* In this spot, you are avoiding the spine and the lowest rib (those bones would decrease penetration.) Depending on how wide your spine is, you will be about 5–7 cm (2 ½ inches) to the left of midline.

Treat the *right* side, the same distance from midline as the left. On the right, however, you want to treat higher by one or two ribs.

Treat between the ribs for better penetration. If you are not sure, do not fret, you will be close enough.

With a 5-milliwatt laser pointer, treat each side for one minute. Move the tip around in a 2 x 2-cm area, the size of a medium-large coin.

Illuminate each side for five seconds with a 100-milliwatt coherent light.

You will be treating the adrenal glands once every seven days until you get your first "blood pressure reaction." This is monitored by your blood pressure results. It will probably come with the first treatment, unless your adrenal glands are severely exhausted. Keep it light, and keep it consistent.

You should see a blood pressure rise of *one to nine* points, on either the systolic (higher number), diastolic, or both. It should only last for a day or two. We are looking for that "blood pressure response." The lower, the better. It does not take much of a dose to go higher, especially if your adrenals are getting stronger. Keep monitoring your blood pressure daily, and keep a record on a calendar. If your blood pressure ever goes more than ten points higher, that is a *"blood pressure over-reaction."* Wait for two weeks after the pressure goes down, and try with ¼ to ½ a dose. It all depends on the reaction.

21

MISCELLANEOUS†

BEDSORES[278], DIABETIC ULCERS†

(MEDIUM CONFIDENCE/LOW RISK)

The research conflicts on this. I noticed that one of the recent negative studies used a 904-nanometer wavelength invisible red coherent light.[279] Generally speaking, the higher the number, the deeper the penetration and the more "shocking" the light to the tissues. Ulcers and skin problems seem to "like" visible red light. Perhaps it is more soothing/"buffered." They are now using simple banks of many bright red or invisible red LEDs in physical therapy clinics to treat diabetic neuropathy and leg ulcers. They used them in the laser therapy clinic in Toronto mentioned earlier. You have that color in your laser pointer (630–660 nanometers). It is a classic red color range used in coherent light treatment and research. In comparing the effectiveness of normal LEDs to laser diodes, a great discussion among experts that points to coherent light being superior can be found at www.laser.nu/lllt/laser_discussion.htm.

Cover the front half of the flat-tipped 5-milliwatt laser pointer with thin plastic wrap. Illuminate the bedsore or ulcer, including the good tissue around the perimeter according to the DelRe Protocols in Chapter 6. Do the same if using a 100-milliwatt coherent light.

SINUSES[†]

(HIGH CONFIDENCE/MODERATE RISK FOR EYE EXPOSURE)

You should use an assistant for this. The laser pointer will be pointed towards your face until it touches the skin. It will be used around your eyes. *There is a risk of hitting the button before the laser pointer is against the skin.* When it is against the skin, you are safe. Assuming you stay away from dairy products, this should work well for you in the short term. Since allergy symptoms come and go, you will probably need to do it again for the next flare-up.

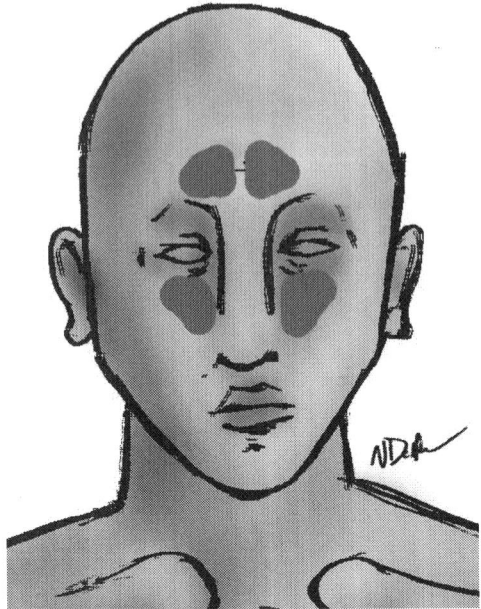

Figure 22: Sinuses

Please refer to figure 22. You must plan to *be careful not to get the laser in your eyes! Any time you are working on your face, please be very careful. Do not turn the laser pointer on until it is firmly against the skin.*

With a 5-milliwatt flat-tipped red laser pointer, illuminate the areas shown in red in Figure 21 for thirty seconds each. If using a 100-milliwatt coherent light, illuminate for seven seconds each.

Limit treatments to once per week for four weeks as long as it is improving and once monthly at the most after that. I do not think it is a good idea to treat more often than that in the long run. Hopefully the sinuses will see some permanent improvement.

Please remember to keep the beam out of your eyes!

HEMORRHOIDS[†]

(HIGH CONFIDENCE/LOW RISK)

There are clinics now that specialize in laser therapy for hemorrhoids. They use higher power lasers, and over a series of treatments, they shrink the hemorrhoids. The method I suggest here is *not the same.*

There has been evidence that you do not need high power lasers that coagulate your protein to shrink hemorrhoids. You can get relief with low power coherent light. It also helps to rejuvenate the area and its weakened blood vessels.[281]

To treat the anal area, cover your laser pointer with thin plastic wrap.

Place a *flat-tipped* 5-milliwatt laser pointer *directly against the skin* that forms *the anus, which is made of a "donut" of muscle*, called a sphincter muscle.

Part 1. Illuminate the outer area for one to two minutes, slowly moving around the "donut."

Part 2. Then, *slightly* insert the laser pointer <u>only</u> about *0.5 cm* (¼ inch). Tilt the flat-tipped 5-milliwatt red laser pointer 45° and rotate it slowly around so that you illuminate the entire circumference of the "donut" there from the inside. Do this for sixty seconds.

Illuminate for twenty seconds if you are using a 100-milliwatt coherent light. You do not need to insert at all, because the 100-milliwatt coherent light will penetrate 4 cm (1½ inches).

You may notice some irritation in the beginning, which will recede over the next twenty-four hours. You can use the DelRe Protocols for this from the start: use two treatments per week to start, and then cut back to once per week when you have reached MMI. Refer to Chapter 6. Wash your hands when you are done, and discard the plastic wrap in a sanitary manner. You may want to wipe down the laser with 70% alcohol afterwards to be sure. Use just a little alcohol on a cloth because you do not want the laser to get wet.

Disk Bulge[†]

(Moderate Confidence/Low Risk)

I could not find any research regarding the use of low-level coherent light and disk damage. However, in my office, I have used it on some very bad, chronic cases of disk bulging/herniation. These are severe cases that are unlikely to regain a pain-free/sciatica free status, at least for a few years of healing or until they decide in favor of surgery. In case you are wondering, I do not "crack them" repeatedly. Although the coherent light therapy has not "made it go away," we have seen some positive changes in healing and pain level by treating the L5-S1 vertebral disk. Be careful here! There is likely to be some osteoarthritis present, and you do not want a flare-up.

What to do to avoid a flare-up is to use a *5-milliwatt red laser pointer and limit the illumination period to sixty seconds on one spot in the center of the spine at the belt line. Limit illumination time to only 3–4 seconds, and no more, with a 100-milliwatt coherent light. In addition, this treatment should be limited to once per week at the beginning, cutting back once MMI is achieved.* Just probe around there, and you will find the *most tender spot* in the midline of the spine, near or below the belt line. That is where you place the laser pointer.

I believe the positive results revealed are because we are causing the disk itself to go through minor healing cascades each time I illuminate the area of the L5-S1 disk. There may be *annular tearing* in the disk. The disk has rings like an onion. Those rings can split from one another, and this causes pain and invites further disk degeneration. I sometimes think that the coherent light may be helping them to heal by "gluing" back together.

Tonsillitis[† 288,289]

(High Confidence, Low Risk)

Your doctor has informed you that milk products can contribute to, or even cause, tonsillitis, right? It does not matter if you are allergic or not, if you have tonsillitis, please stay away from milk products. Do not consume milk products until you are well clear of the tonsillitis.

Do not illuminate the tonsils if you have a <u>fever</u>. Be aware that the thyroid may become stimulated and will probably increase a fever if you have one.

You will need to use a mirror, and with mirrors and lasers, you always need to be extra careful. With a 5-milliwatt flat-tipped red laser pointer, illuminate each tonsil for sixty seconds. Since you will not be able to put the tip of the laser against the tissue, you should get as close as reasonable and keep the light on one spot for fifteen seconds before moving to the next spot (four spots per tonsil.) If using a 100-milliwatt coherent light, it will cover the entire tonsil. Illuminate for ten seconds on each tonsil. This treatment will not affect a bacterial infection significantly, but it will stimulate the tonsils to work more effectively and heal.

Treat once per week, cutting back to two-week intervals once or twice when the tonsillitis episode is over. For treatment of chronic cases, illuminate the tonsils once a month for 4–6 months.

As with all glands, we must assume that we could "burn them out," exhausting them, either permanently or temporarily. Please do not exceed the above recommended treatment protocols.

"There are grades of vanity, there are only grades of ability in concealing it."
William Shakespeare
The Tragedy of King Richard the Second (York at II, i)

"O the days of youth,
So far away,
yet so near,
If only we let them in".[290]

When it comes down to it, any remedy is a temporary fix. If I believe in a spiritual Creator, then I must admit that sickness is a spiritual matter between laws of eternity and my soul, played out in time. Yep, I said that. That is a hard responsibility to accept. In my opinion, there is only one permanent healing, and that is the resurrection, in whatever form it comes.

It is a vanity for a twenty-year-old to use this information, risking his future health for an idea of personal beauty.

It is less of a vanity for a testosterone-deficient male to use this information to improve his muscle tone and quality of life.

It is no vanity to use this information to bring the feeling back to the feet in a diabetic, pain relief to the rheumatoid arthritic, a warmer heart, or a sane new mother.

It is a vanity to ruin one's health in order to lose weight to unnatural thinness.

It is no vanity to improve one's health by following conservative instructions to escape obesity.

111

It is no vanity always to reach for the optimism of youth. It takes humility. If we can do this perfectly, forgiving the failures, we will find we have better health in general.

IMPORTANT!

.

Please leave a review at: http://www.amazon.com/
Rejuvenation-Increase-Vitality-Energy-Healing/product-
reviews/0615263275

.

If you don't want to type all that out, just go to amazon.com
and look up "Rejuvenation" and it will take you to where
you will see a button to leave reviews.

.

Every review a reader gives will decrease the time necessary
for this information to be used by the ones we trust to know
everything: Doctors!

INDEX

113

Bursitis x, xiv, 10, 16, 23, 26, 37, 38, 39, 120, 127
  olecranon 38
  prepatellar 37
  subacromial 38
  trochanteric 38

C

Calcium 42, 91
  bank 91
Calcium/Magnesium 56, 65, 90, 91
Cancer 2, 11, 13, 16, 50, 55, 67, 72, 119, 124, 126, 128, 129
Cancerous 72
Capillaries 27, 44, 86
Cardiac 26, 119, 125, 129
Carotid arteries 71
Carpal bones 48
  carpal tunnel arch 48
Carpal tunnel arch 48
Carpal Tunnel Syndrome xiv, 47, 48, 126, 128
Caucasian 21
Celiac Plexus 81
Cell
  cell membrane resistance 9
  cell wall 9
  exercising 12, 25, 66
cells xv, 8, 9, 10, 12, 17, 25, 26, 27, 28, 50, 54, 63, 66, 67, 72, 75, 86, 87, 91, 121, 123, 124, 125, 129, 134
Cellular respiration 10, 25, 66
Cervix xiv
Chakra 51, 74
  root 48, 51, 85, 86, 87, 88, 89
Cheap gradeî of tissue 27
Checks and balances 93
Cheek 88, 89
Children 13, 16, 17, 18, 19, 23, 133, 134
Chiropractic 56
Chiropractor xv, 3, 48, 56, 62, 90
Chiropractors xvi, 56
chloroblasts 8
cholesterol 8, 43, 67, 119, 127
Cholesterol
  decrease 8, 43, 67, 119, 127
  LDL 8, 127
Chronic xvi, 5, 6, 7, 9, 10, 17, 18, 27, 55, 98, 108, 109, 120, 125, 126, 127, 128, 129, 132
Chronic condition 120
Coffee 95
Coherent light xiv, xv, 10, 11, 50, 55, 67, 74, 78, 86
Coherent light techniques 6
Coherent light therapy xiv, xvii, 1, 2, 3, 5, 6, 7, 8, 10, 11, 12, 15, 16, 17, 18, 19, 44, 47, 48, 49, 50, 52, 55, 56, 59, 61, 63, 66, 67, 73, 74, 75, 78, 79, 81, 86, 87, 89, 94, 96, 108
Cold lasers 10

114

Collagen 26
Collateral repair 87
Composite
  poisoning 86
Consistency xv, 11, 65
Corticosterone 60
Cortisol 95, 96
Cortisone 42, 96
Cramping 56
Cramps 56, 57, 91

D

Dark skin 21
de-bunking xiv
Deeper tissues 6
DelRe Protocols 15, 39, 42, 44, 48, 55, 72, 74, 88, 89, 107
Dental
  bone problems xiv, 2, 89, 120, 133
  Implant Success 89
  pulp 85, 87, 90, 132, 133
  root canal 88
  TMJ Syndrome xiv, 90, 92
Dentist 85, 86, 87, 89, 90
Dentistry 85
Department of Laser Therapy 10
Detached retina, 44
Diabetes 1, 43, 44, 96, 119, 122, 124, 125, 126, 127
Diabetic
  Diabetic Neuropathy 44
  neuropathy 16
  retinopathy 44
  ulcers 25, 123
Diabetic retinopathy 44
Diabetics 8, 43, 45
Direct Treatment 43
  Heart 74
Disk 90, 108
Disk Bulge 108
DNA 11, 67, 129
Dosage xv, 5, 13, 16, 21, 22, 23, 42, 45, 51, 61, 71, 78, 88
Duodenal ulcers 81, 132
Duodenum 81, 82
Dysplasia 50

E

Eggs 56
Elbow x, 38
  bursitis x, 38
Electronic circuits 53
Electronics technician 93
Elemental nutrient 91
Endocrinology 2
Endodontic surgery
  pain with 85

115

REFERENCES

†The information provided in this book should not be construed as personal medical advice or instruction as it is for educational purposes only. This book, the publisher and the author make no claims about prevention of, diagnosis for, treatments for, or cures of specific ailments or diseases. If you have a medical condition you should seek professional medical advice from your physician. This book is not a substitute for the medical advice, counsel and treatment of medical professionals. This book should be used only as reference tool to assist you. No action should be taken by anyone based solely on the contents of this book. While this book represents helpful information it is impossible for any book to be able to identify you or your loved one's medical problems or to make specific medical recommendations for you or a loved one. One's medical needs, specific treatment options, specific medical care, and specific treatment fees can only be determined after clinically seeking the recommendation of licensed medical professionals. Readers should consult appropriate health professionals on any matter relating to their health and well-being. The information and opinions provided here are believed to be accurate and sound, based on the best judgment available to the author, but readers who fail to consult appropriate health authorities assume the risk of any injuries. The author or publisher is not responsible for errors or omissions.

1   Photoengineering of Tissue Repair in Skeletal and Cardiac Muscles Apr 2006, Vol. 24, No. 2: 111-120 , Photomedicine and Laser Surgery. Uri Oron, Ph.D.Department of Zoology, The George S. Wise Faculty of Life Sciences, Tel-Aviv University, Tel-Aviv, Israel.

2   Efficacy of low level laser therapy on neurosensory recovery after injury to the inferior alveolar nerve. Tuncer Ozen, Kaan Orhan, Ilker Gorur, Adnan Ozturk. Head Face Med. 2006; 2: 3.

3   Low power laser therapy of shoulder tendinitis. England S, Farrell AJ, Coppock JS, Struthers G, Bacon PA. Scand J Rheumatol. 1989;18(6):427–431.

4   Using lasers in diabetic wound healing.Diabetes Forney R, Mauro T. Technol Ther. 1999 Summer;1(2):189-92.

5   Use Of Infrared Laser Therapy In Patients With Ischemic Heart Disease Associated With Diabetes Mellitus Type 2   Zin'kovskaia TM, Zavrazhnykh LA, Golubev AD.

6   The use of laser radiation for correcting cholesterol metabolic disorders in gallstones. Grubnik, V.V. et al. Klin Khir. 1993; (11): 18-20.

7   Possibilities of the treatment of biliary dyskinesia by laser irradiation. L P Vorob'ev, L M Salova, V M Meshkov, G V Chubarov Klin Med (Mosk) May 1990 (Vol. 68, Issue 5, Pages 110-5)

8   Complex treatment of non-specific ulcerative colitis with low-level HeNe laser used transanally. Moscow Abstracts. Dubinkin V A, Mimrikova E G.

9   Immunological and microbiological aspects of low intensity laser effect on the factors of local immunity of the reproductive tract in women with chlamydia infection. Dolgushin II, Gizinger OA, Telesheva LF. Zh Mikrobiol Epidemiol Immunobiol. 2006 Jul-Aug;(4):105-9.

10 Clinical Application Of GaAlAs 830 nm Diode Laser (LLLT) In Treatment Of Rheumatoid Arthritis Kanji Aoada, Yasutaka Ytani, Akira Sakawa and Akira Shimazu - Department of Orthopaedic Surgery, Osaka City University Medical School, Japan

11 Efficacy Of Low Power Laser Therapy In Fibromyalgia: A Single-Blind, Placebo-Controlled Trial. Gur A, Karakoc M, Nas K, Cevik R, Sarac J, Demir E.Physical Medicine and Rehabilitation, School of Medicine, Dicle University, Diyarbakir, Turkey. Lasers Med Sci. 2002;17(1):57-61.

12 A systematic review of common conservative therapies for arm lymphoedema secondary to breast cancer treatment. Moseley AL, Carati CJ, Piller NB. Ann Oncol. 2007 Apr;18(4):639-46. Epub 2006 Oct 3.

13   FDA Clears Laser-Based Lymphedema Therapy   Oct/Nov 2006

News Release; The Food and Drug Administration (FDA) has cleared the RianCorp LTU-904 low-level laser device as an additional weapon in the battle for better quality of life for people living with lymphedema. After a four-year evaluation, the FDA cleared the LTU-904 for use as part of a therapy regime to treat post-mastectomy lymphedema. FDA clearance permits therapists throughout the USA to use the handheld battery-powered device to treat a patient with lymphedema.

Lymphedema, which affects as many as 30% of post-mastectomy patients, is a chronic condition that impacts about three million people in the US. It causes a person's limb to enlarge because lymphatic fluid does not drain from the limb after the lymphatic system is compromised following breast surgery. Since 2000, therapists in Australia have used the LTU-904 to treat patients with lymphedema. The FDA noted that the RianCorp LTU-904 laser therapy unit has been evaluated in a placebo-controlled double blind clinical trial of post-mastectomy lymphedema patients in Australia.

Conducted under ethics committee approval by Flinders University, the study is the world's only randomised double blind study of a physical treatment for post-mastectomy lymphedema. The Flinders research team is internationally recognised in the area of lymphology and lymphedema treatment. More half of patients receiving LTU-904-based therapy experienced a reduction in ECF The trial showed that 52% of patients experienced a clinically significant decrease in ECF (Extra Cellular Fluid) after six weeks of laser treatment. In contract only 19% of placebo patients experienced the same result. The trial's results were published in the highly-regarded peer-reviewed journal "Cancer".

The Flinders study enrolled 64 post-mastectomy patients with at least 200 ml difference between their arms. A summary of the clinical trial results is available at http://www.riancorp.com/.

The LTU-904 Laser Therapy Unit is a non-thermal device that delivers a controlled series of 200 ns bursts of pulses of 904nm laser beam. The near-infrared beam is invisible to the human eye. The FDA noted that investigators observed no adverse effects from the laser treatments and the study demonstrated the LTU-904 functioned as intended in all treatments of post-mastectomy lymphedema.

14   Effect Of Low-Power Laser Irradiation On Bony Implant Sites. Dortbudak O, Haas R, Mailath-Pokorny G. Dept of Oral Surgery, Dental School, University of Vienna, Austria. Clin Oral Implants Res. 2002 Jun;13(3):288-92.

15   Low intensity laser application in temporomandibular disorders: a phase I double-blind study. Mazzetto MO, Carrasco TG, Bidinelo EF, de Andrade Pizzo RC, Mazzetto RG. Cranio. 2007 Jul;25(3):186-92.

16   Carpal tunnel syndrome pain treated with low-level laser and microamperes transcutaneous electric nerve stimulation: A controlled study. Aeser MA, Hahn KA, Lieberman BE, Branco KF. Dept of Neurology, Boston University School of Medicine, Psychology Research Service, MA, USA.

17   "Low level laser therapy in shoulder tendinitis/bursitis, epicondylalgia and ankle sprain" Jan M Bjordal, J.M. 1997, Division of Physiotherapy Science, University of Bergen Also in Physical Therapy Reviews. 1998; 3: 121-132.

18   A randomised, placebo controlled trial of low level laser therapy for activated Achilles tendinitis with microdialysis measurement of peritendinous prostaglandin E2 concentrations.Bjordal, J M; Lopes-Martins, R A B; Iversen, V V British Journal of Sports Medicine. 40(1):76-80, January 2006.

19   Low level laser therapy for tendinopathy. Evidence of a dose-response pattern. BJORDAL Jan Magnus; COUPPE Christian; LJUNGGREN Anne Elisabeth. Physical therapy reviews 2001, vol. 6, no2, pp. 91-99

20   Quantitative and qualitative changes of the seminiferous epithelium induced by Ga. Al. As. (830 nm) laser radiation. Taha MF, Valojerdi MR. Lasers Surg Med. 2004;34(4):352-9.

21   Experimental photodynamic laser therapy for rheumatoid arthritis with a second generation photosensitizer. Hendrich C, Huttmann G, Vispo-Seara JL, Houserek S, Siebert WE. Knee Surg Sports Traumatol Arthrosc. 2000;8(3):190-4.

22   NI. Low-intensity laser radiation in preoperative preparation of patients with benign prostatic hyperplasia. Neimark AI, Muzalevskaia Urologiia. 2000 Jan-Feb;(1):11-5.

23   Low-Intensity Laser Therapy is an Effective Treatment for Recurrent Herpes Simplex Infection. Results from a Randomized Double-Blind Placebo-Controlled StudyAndreas Schindl and Reinhard Neumann Journal of Investigative Dermatology (1999) 113, 221–223; doi:10.1046/j.1523-1747.1999.00684.x

24   Infrared laser therapy for ischemic stroke: a new treatment strategy: results of the NeuroThera Effectiveness and Safety Trial-1 (NEST-1). Lampl Y, Zivin JA, Fisher M, Lew R, Welin L, Dahlof B, Borenstein P, Andersson B, Perez J, Caparo C, Ilic S, Oron U, Stroke; Jun 2007 (Vol. 38, Issue 6, Pages 1843-9)

25   Transcranial infrared laser therapy improves clinical rating scores after embolic strokes in rabbits. Paul A Lapchak, Jiandong Wei, Justin A Zivin, Stroke; Aug 2004 (Vol. 35, Issue 8, Pages 1985-8)

26   Low-level laser therapy applied transcranially to rats after induction of stroke significantly reduces long-term neurological deficits. Amir Oron, Uri Oron, Jieli Chen, Anda Eilam, Chunling Zhang, Menachem Sadeh, Yair Lampl, Jackson Streeter, Luis DeTaboada, Michael Chopp, Stroke; Oct 2006 (Vol. 37, Issue 10, Pages 2620-4)

27   From The Benefits of Nutritional Supplements, Copyright © 2002 Council for Responsible Nutrition (CRN) Intake of antioxidants among American cardiologists. Mehta J. Am J Cardiol 1997; 79:1558-1560.

28   Is US Health Really the Best in the World? B Starfield. JAMA. 2000;284:483-485.

29   Exocrine and endocrine testicular function during the treatment of experimental orchitis and non-specific orchoepididymitis by low-energy laser radiation. Leonid L. Reznikov Proceedings of SPIE -- Volume 2391; Laser-Tissue Interaction VI, Steven L. Jacques, Editor, May 1995, pp. 641-647; Univ. of Colorado Health Sciences Ctr. (USA) Ludmila S. Pupkova St. Petersburg Scientific Research Ctr. (Russia) H. Bell Embryology Associates, Inc. (USA) Alexander G. Murzin St. Petersburg Scientific Research Ctr. (Russia)

30   Exposure of the volunteers' skin to visible polarized light is accompanied by increase in growth promoting activity of the circulation blood for human cells in vitro: Dependence of the effect on cell type. K.A.Samoilova, M.I.Blinova, N.M.Yudintzeva, N.V.Kalmykova, E.V.Kuz'minych *D.I.Sokolov, O.I.Zubanova, K.D.Obolenskaya Institute of Cytology of the Russian Academy of Sciences, Institute for Experimental Medicine of the Russian Academy of Medical Sciences, St.Petersburg, Russia

31   Effects of low-power laser radiation on mice immunity. E G Novoselova, O V Glushkova, D A Cherenkov, V M Chudnovsky, E E Fesenko. Photodermatol Photoimmunol Photomed.Feb 2006 (Vol. 22, Issue 1, Pages 33-8)

32   Low level laser irradiation shows no histopathological effect on myogenic satellite cells in tissue culture. G. Shefer, M. Cullen, O. Halevi and U. Oron. Laser Therapy 11:114-117.2001.

33   To Examine the Adverse Photothermal Effects of Extended Dosage Laser Therapyin Vivo on the Skin and Subcutaneous Tissue in the Rat Model, Sasaki , K., et al. Laser Therapy, Vol. 4, No.2, Apr-Jun 1992

34   A Study on the Possible Haemorrhagic Effects of Extended Infrared Diode Laser Irradiation on the Encapsulated and Exposed Synovial Membrane Articular Tissue of the Rat. Calderhead ,R. G. et al. Laser Therapy, Vol. 4, No. 2, Apr.-Jun. 1992

35   Low-energy helium neon laser treatment of thumb osteoarthritis. Basford JR, Sheffield CG, Mair SD, Ilstrup DM. Arch Phys Med Rehabil. 1987 Nov;68(11):794–797.

36   Efficacy of different therapy regimes of low-power laser in painful osteoarthritis of the knee: a double-blind and randomized-controlled trial. Gur A, Cosut A, Sarac AJ, Cevik R, Nas K, Uyar ALasers Surg Med. 2003 (Vol. 33, Issue 5, Pages 330-8)

37   Low power laser treatment in patients with knee osteoarthritis.HHTascioglu F, Armagan O, Tabak Y, Corapci I, Oner C.HH Swiss Med Wkly. 2004 May 1;134(17-18):254-8.

38   Low power Ga-Al-As laser treatment of painful osteoarthritis of the knee. A double-blind placebo-controlled study.Bulow PM, Jensen H, Danneskiold-Samsoe B.   Scand J Rehabil Med. 1994 Sep;26(3):155-9.

39   Experimental photodynamic laser therapy for rheumatoid arthritis with a second generation photosensitizer.Hendrich C, Huttmann G, Vispo-Seara JL, Houserek S, Siebert WE. Knee Surg Sports Traumatol Arthrosc. 2000;8(3):190-4.

40   Clinical Application Of GaAlAs 830 nm Diode Laser (LLLT) In Treatment Of Rheumatoid Arthritis Kanji Aoada, Yasutaka Ytani, Akira Sakawa and Akira Shimazu - Department of Orthopaedic Surgery, Osaka City University Medical School, Japan

41   Evaluation of Acute Pain Relief Effects of Low-Power Laser Therapy on Rheumatoid Arthritis by Thermography. Obata, J., et al, Laser Therapy (1990), 2:28

42   Laser Irradiation of the Blood. Gasparyan, Levon, Laser Partner and Laser World. 10.1.2003

43   Role of the circulating blood in initiation of therapeutic effects of visible light.K.A.Samoilova Institute of Cytology of the Russian Academy of Sciences, St. Petersburg, Russia The Journal of Laser Therapy. Millennium Edition Volumes 12 & 13

44   Evidence that Alzheimer's disease is a micro-vascular disorder: the role of constitutive nitric oxide. JC de la Torre, GB Stefano. Brain Research Reviews Volume 34, Issue 3, December 2000, Pages 119-136

45   Low power laser protects human erythrocytes In an In vitro model of artificial heart-lung machines. Artificial Organs. 2000; 24 (11): 870-3. Itoh-T, Murakami-H, Orihashi-K et al.

46   Low-intensity laser therapy in pediatric oncology. Voprosy Onkologii; 2000, 46 (4): 459-61. Balakirev-S-A, Gusev-L-I, Kazanova-M et al.

47   In vitro effects of low-level laser irradiation at 660 nm on peripheral blood lymphocytes. Lasers in surgery and medicine 2000, 27 (3): 255-261. Stadler I, Evans-R, Kolb B, Naim-J-Oet al.

48   Dynamics of hyperlipidemia and peripheral blood flow in patients with diabetes mellitus after the course of combined laser therapy in ambulatory-polyclinic conditions. Oprysko T V et al.

49   Effect of Multiple Exposures of Low-Level Laser Therapy on the Cellular Responses of Wounded Human Skin Fibroblasts. Denise Hawkins, Heidi Abrahamse. Photomedicine and Laser Surgery. December 1, 2006, 24(6): 705-714. doi:10.1089/pho.2006.24.705.

50   Molecular mechanisms of therapeutic effects of low-intensity laser radiation. Karu TI. Laser Life Sciences, 1988;2:53.

51   Membranotropic effect of the low energy laser irradiation of the blood. A A Spasov, V V Nedogoda, O V Ostrovskiï, K Konan Biull Eksp Biol Med. Oct 1998 (Vol. 126, Issue 10, Pages 412-5)

52   Histological studies on the rheumatoid synovial membrane irradiated with a low energy laser Dr. A. Amano, MD, K. Miyagi, MD, T. Azuma, MD, Y. Ishihara, MD, S. Katsube, MD, I. Aoyama, MD, I. Saito, MD Department of Internal Medicine, Toyoko Hospital, St. Marianna University, Kanagawa, Japan

53   Mechanisms of low-power laser light action on cellular level. Karu, Tiina I. (Laser Technology Research Ctr.)Proc. SPIE Vol. 4159, p. 1-17

54   Effect of red and near-infrared laser light on adenosine triphosphate (ATP) in the luciferine–luciferase reaction. Amat A, Rigau J, Nicolaua R, Aalders M et al. Journal of Photochemistry and Photobiology A: Chemistry. 2004; 168 (1-2): 59-65.

55   Ga-As (808 nm) Laser Irradiation Enhances ATP Production in Human Neuronal Cells in Culture; U. Oron, Ph.D. et al. Photomedicine and Laser Surgery; Jun 2007, Vol. 25, No. 3 : 180 -182.

56   The effect of scatter laser photocoagulation on intravitreal levels of growth factors in the miniature pig. M Xiao, A Khaliq, P Moriarty, D McLeod, J Cranley, M Boulton Curr Eye Res Sep 1996 (Vol. 15, Issue 9, Pages 923-31)

57   Treatment of exp induced transient cerebral ischemia with low energy laser inhibits nitric oxide synthase activity and up-regulates the expression of transforming growth factor-beta 1. Mason C P Leung, Samuel C L Lo, Flora K W Siu, K-F So Lasers Surg Med 2002 (Vol. 31, Issue 4, Pages 283-8)

58   A study on expression of basic fibroblast growth factors in periodontal tissue following orthodontic tooth movement associated with low power laser irradiation Xianchun Zhu, Yuanping Chen, Xinhua Sun Hua Xi Kou Qiang Yi Xue Za Zhi Jun 2002 (Vol. 20, Issue 3, Pages 166-8)

59   Low-energy laser irradiation increases endothelial cell proliferation, migration, and eNOS gene expression possibly via PI3K signal pathway. Chen CH, Hung HS, Hsu SH. Lasers Surg Med. 2008 Jan;40(1):46-54.

60   Augmentation of the expression of proangiogenic genes in cardiomyocytes with low dose laser irradiation in vitro. Ashwani Khanna, Latha Raja Shankar, Michael H. Keelan, Ran Kornowski, Martin Leon, Jeffrey Moses, Nicholas Kipshidze Cardiovascular Radiation Medicine July 1999 (Vol. 1, Issue 3, Pages 265-269)

61   Low-level laser therapy accelerates collateral circulation and enhances micro circulation. F R Mohammed Ihsan Photomed Laser Surg Jun 2005 (Vol. 23, Issue 3, Pages 289-94)

62   Effect of laser on hair growth of mice". Mester, E.; Szende, B., and Tota, J.G. (1967). Kiserl Orvostud 19: 628-631.

63 An investigation of the cytotoxic and mutagenic potential of low intensity laser irradiation in Friend erythroleukaemia cells. I D Logan, P G McKenna, Y A Barnett Mutat Res Jul 1995 (Vol. 347, Issue 2, Pages 67-71)

64 Low level laser irradiation shows no histopathological effect on myogenic satellite cells in tissue culture. G. Shefer, M. Cullen, O. Halevi and U. Oron Laser Therapy 11:114-117.2001.

65 To Examine the Adverse Photothermal Effects of Extended Dosage Laser Therapyin Vivo on the Skin and Subcutaneous Tissue in the Rat Model. Sasaki , K., et al, Laser Therapy, Vol. 4, No.2, Apr-Jun 1992

66 A Study on the Possible Haemorrhagic Effects of Extended Infrared DiodeLaser Irradiation on the Encapsulated and Exposed Synovial Membrane Articular Tissue ofthe Rat. Calderhead ,R. G. et al, Laser Therapy, Vol. 4, No. 2, Apr.-Jun. 1992

67 Low level laser irradiation stimulates mitochondrial membrane potential and disperses subnuclear promyelocytic leukemia protein. Gavish L; Asher Y; Becker Y; Kleinman Y Lasers in surgery and medicine; VOL: 35 (5); p. 369-76 /2004/ Department of Molecular Virology, The Faculty of Medicine, The Hebrew University of Jerusalem

68 Effect of a diode laser on wound healing by using diabetic and nondiabetic mice. Kawalec J S, Hetherington J, Pfennigwerth C et al. Journal of Foot and Ankle Surgery. 2004; 43 (4): 214-220.

69 Dose and wavelength of laser light have influence on the repair of cutaneous wounds. Mendez TM; Pinheiro AL; Pacheco MT; Nascimento PM; Ramalho LM Journal of clinical laser medicine & surgery; VOL: 22 (1); p. 19-25 /200402

70 Low-level laser therapy (LLLT) efficacy in post-operative wounds. Herascu N; Velciu B; Calin M; Savastru D; Talianu C Photomedicine and laser surgery; VOL: 23 (1); p. 70-3 /200502/

71 Effects of a therapeutic laser on the ultrastructural morphology of repairing medial collateral ligament in a rat model. Fung DT; Ng GY; Leung MC; Tay DK Lasers in surgery and medicine; VOL: 32 (4); p. 286-93 /2003/

72 Effect of low-level laser treatment on neurosensory deficits subsequent to saggittal split ramus osteotomy Shelley M. Khullar, Babak Emami, Anders Westermark, Hans R. Haanaes Oral Surgery, Oral Medicine, Oral Pathology, Oral Radiology and Endodontics Aug 1996 (Vol. 82, Issue 2, Pages 132-138)

73 Effects of low-power laser radiation on mice immunity. E G Novoselova, O V Glushkova, D A Cherenkov, V M Chudnovsky, E E Fesenko. Photodermatol Photoimmunol Photomed Feb 2006 (Vol. 22, Issue 1, Pages 33-8)

74 Effects of GaAs-lasers combined with radiotherapy on murine sarcoma deoends on tumor size. Saldo I et al. Lasers in Surgery and Medicine. 1989; Suppl 1:40.

75 Investigations on the influence of low level laser diode laser irradiation of the growth of experimental tumors. Mikhailov V A et al.Laser Therapy. 1993;5(1):33-38

76 . Dose related immunological and morphological changes observed in rats with Walker - 256 carcinosarcoma after photodynamic theraoy: a controlled study. Dima F V et al. Laser Therapy. 1991;3(3):159-168

77 Oncological aspects of the biological action of low level laser radiation. II. Investigation of primary mechanisms and clinical applications Ivanov A V, Zakharov S D. SPIE proceedings series Laser use in oncology II : (selected papers on laser use in oncology 1997-1999) 2000, vol. 4059, pp. 101-111, [Note(s) : VIII, 230 p.,] (28 ref.) ISBN 0-8194-3688-7

78 A Study on the Possible Haemorrhagic Effects of Extended Infrared DiodeLaser Irradiation on the Encapsulated and Exposed Synovial Membrane Articular Tissue ofthe Rat. Calderhead ,R. G. et al. Laser Therapy, Vol. 4, No. 2, Apr.-Jun. 1992

79 Effect of In-Ga-Al-P diode laser irradiation on angiogenesis in partial ruptures of Achilles tendon in rats. Salate AC, Barbosa G, Gaspar P, Koeke PU, Parizotto NA, Benze BG, Foschiani D. Photomed Laser Surg. 2005 Oct;23(5):470-5.

80 Laser treatment for stroke. Neurother. 2007 Aug;7(8):961-5. Lampl Y. Edith Wolfson Medical Center, Department of Neurology, Holon, Israel.

81 Low-energy laser irradiation increases endothelial cell proliferation, migration, and eNOS gene expression possibly via PI3K signal pathway. Chen CH, Hung HS, Hsu SH. Lasers Surg Med. 2008 Jan;40(1):46-54.

82 Effects Of Pulse Frequency Of Low-Level Laser Therapy (LLLT) On Bone Nodule Formation In Rat Calvarial Cells Ueda Y, Shimizu N. Department of Orthodontics, Nihon University School of Dentistry at Matsudo Chiba, Japan. J Clin Laser Med Surg. 2003 Oct;21(5):271-7.

83   Effect Of 830-Nm Laser Light On The Repair Of Bone Defects Grafted With Inorganic Bovine Bone And Decalcified Cortical Osseous Membrane. Barbos Pinheiro AL, Limeira Junior Fde A, Marquez Gerbi ME, Pedreira Ramalho LM, Marzola C, Carneiro Ponzi EA, Oliveira Soares A, Bandeira De Carvalho LC, Vieira Lima HC, Oliveira Goncalves T. Laser Center, School of Dentistry, Federal University of Bahia, Salvador, Brazil. J Clin Laser Med Surg. 2003 Dec;21(6):383-8.

84   Low-Power Laser Irradiation Improves Histomorphometrical Parameters and Bone Matrix Organization During Tibia Wound Healing In Rats Garavello-Freitas I, Baranauskas V, Joazeiro PP, Padovani CR, Dal Pai-Silva M, da Cruz-Hofling MA. J Photochem Photobiol B. 2003 May-Jun;70(2):81-9.

85   Bone Repair Of The Periapical Lesions Treated Or Not With Low Intensity Laser (Wavelenght=904 nm) Laser Surg Med. Issue 2002. Sousa G R, Ribeiro M S, Groth E B.

86   Computerized Morphometric Assessment Of The Effect Of Low-Level Laser Therapy On Bone Repair: An Experimental Animal Study Silva Júnior AN, Pinheiro AL, Oliveira MG, Weismann R, Ramalho LM, Nicolau RA. J Clin Laser Med Surg. 2002; 20: 83-87

87   Laser Therapy Plays A Role In Bone Healing   Lasers Surg Med. 1998; 22: 97-102. Luger et al. studied the effect of HeNe laser on the healing of tibial bone fractures in rats.

88   Effect Of Low-Power Gaalas Laser (660 Nm) On Bone Structure And Cell Activity: An Experimental Animal Study Nicola RA, Jorgetti V, Rigau J, Pacheco MT, dos Reis LM, Zangaro RA. Lasers Med Sci. 2003;18(2):89-94.

89   Photoengineering of Bone Repair Processes Apr 2006, Vol. 24, No. 2: 169-178, Photomedicine and Laser Surgery Dr. Antonio Luiz B. Pinheiro, D.D.S., Ph.D. Laser Center, School of Dentistry, Department of Propedêutica and Clínica Integrada, Universidade Federal da Bahia, Canela Salvador, BA, Salvador, Brazil. Institute for Research and Development, Universidade do Vale do Paraíba, Sˇo José dos Campos, SP, Brazil.Marleny Elizabeth M.M. Gerbi, Ph.D.

90   Photomedicine and Laser Surgery Effect of Laser Therapy on Bone Tissue Submitted to Radiotherapy: Experimental Study in Rats Jun 2007, Vol. 25, No. 3 : 197 -204 Samantha Seara Da Cunha, D.D.S., Ph.D.

91   Low-intensity laser irradiation stimulates bone nodule formation via insulin-like growth factor-I expression in rat calvarial cells. Shimizu N, Mayahara K, Kiyosaki T, Yamaguchi A, Ozawa Y, Abiko Y. Lasers Surg Med. 2007 Jul;39(6):551-9.

92   Laser technology in orthopedics: preliminary study on low power laser therapy to improve the bone-biomaterial interface Guzzardella GA, Torricelli P, Nicoli Aldini N, Giardino R. .Int J Artif Organs. 2001 Dec;24(12):898-902.

93   The efficacy of laser therapy in wound repair: a meta-analysis of the literature. Photomed Laser Surg. 2004 Jun;22(3):241-7. Woodruff LD, Bounkeo JM, Brannon WM, Dawes KS, Barham CD, Waddell DL, Enwemeka CS. Department of Physical Therapy, North Georgia College and State University, Dahlonega, Georgia, USA.

94   Low intensity laser irradiation in the treatment of recalcitrant radiation ulcers in patients with breast cancer--long-term results of 3 cases. Schindl A, Schindl M, Pernerstorfer-Schon H, Mossbacher U, Schindl L. Photodermatol Photoimmunol Photomed. 2000 Feb;16(1):34-7.

95   Healing of gingival incisions by the helium-neon diode laser. Neiburger EJ.Rapid J Mass Dent Soc. 1999 Spring;48(1):8-13, 40.

96   Using lasers in diabetic wound healing. Forney R, Mauro T Diabetes Technol Ther. 1999 Summer;1(2):189-92.

97   Low-Level Laser Therapy Facilitates Superficial Wound Healing in Humans: A Triple-Blind, Sham-Controlled Study J. Ty Hopkins, Todd A. McLoda, Jeff G. Seegmiller, and G. David Baxter J Athl Train. 2004 Jul–Sep; 39(3): 223–229.

98   Photobiomodulation improves cutaneous wound healing in an animal model of type II diabetes. Byrnes KR; Barna L; Chenault VM; Waynant RW; Ilev IK; Longo L; Miracco C; Johnson B; Anders JJ. Photomedicine and laser surgery; VOL: 22 (4); p. 281-90 /200408/ Department of Anatomy, Physiology and Genetics, Uniformed Services University of the Health Sciences, 4301 Jones Bridge Road, Bethesda, MD 20814, USA.

99   Effect of low intensity helium-neon (He-Ne) laser irradiation on diabetic wound healing dynamics. Maiya GA; Kumar P; Rao L Photomedicine and laser surgery; VOL: 23 (2); p. 187-90 /200504/ Department of Physiotherapy, MAHE University, Manipal, India.

100   Wound healing of animal and human body sport and traffic accident injuries using low-level laser therapy treatment: a randomized clinical study of seventy-four patients with control group. Simunovic Z; Ivankovich AD; Depolo A Journal of clinical laser medicine & surgery; VOL: 18 (2); p. 67-73 /200004/ Department of Anesthesiology, La Caritá Medical Center, Laser Center, Locarno, Switzerland.

101   Kawalec J S, Hetherington J, Pfennigwerth C et al. Effect of a diode laser on wound healing by using diabetic and nondiabetic mice. Journal of Foot and Ankle Surgery. 2004; 43 (4): 214-220.

102   Comparison between Wound Healing in Induced Diabetic and Non diabetic Rats after Low-Level Laser Therapy Aug 2006, Vol. 24, No. 4: 474-479 Sylvia Bicalho Rabelo , D.D.S., M.S.D.

103   Histological Assessment of the Effect of Laser Irradiation on Skin Wound Healing in Rats Aug 2006, Vol. 24, No. 4: 480-488   Peter Gál et al. . Department of Medical Biophysics, Faculty of Medicine, Pavol Jozef šafárik University in Ko[scaron]ice, Košice, Slovak Republic.

104   Low level laser irradiation stimulates mitochondrial membrane potential and disperses subnuclear promyelocytic leukemia protein. Gavish L; Asher Y; Becker Y; Kleinman Y Lasers in surgery and medicine; VOL: 35 (5); p. 369-76 /2004/ Department of Molecular Virology, The Faculty of Medicine, The Hebrew University of Jerusalem, Jerusalem, Israel.

105   Low-level laser therapy (LLLT) prevents oxidative stress and reduces fibrosis in rat traumatized Achilles tendon. Fillipin LI, Mauriz JL, Vedovelli K, Moreira AJ, Zettler CG, Lech O, Marroni NP, Gonzalez-Gallego J. Lasers Surg Med. 2005 Oct;37(4):293-300.

106   The effect of low level laser irradiation on implant-tissue interaction. In vivo and in vitro studies. Khadra M. Swed Dent J Suppl. 2005;(172):1-63. Department of Oral Surgery and Oral Medicine, Faculty of Dentistry, University of Oslo, Norway.

107   Lasers Surg Med. 2007 Jul;39(6):551-9. Low-intensity laser irradiation stimulates bone nodule formation via insulin-like growth factor-I expression in rat calvarial cells. Shimizu N, Mayahara K, Kiyosaki T, Yamaguchi A, Ozawa Y, Abiko Y. Department of Orthodontics, Nihon University School of Dentistry, 1-8-13 Kanda, Surugadai, Chiyoda-Ku, Tokyo101-8310, Japan.

108   Photoengineering of Tissue Repair in Skeletal and Cardiac Muscles Apr 2006, Vol. 24, No. 2: 111-120 , Photomedicine and Laser Surgery; Dr. Uri Oron, Ph.D.; Department of Zoology, The George S. Wise Faculty of Life Sciences, Tel-Aviv University, Tel-Aviv, Israel.

109   Mechanisms of low-power laser light action on cellular level. Tiina Karu Institute of Laser and Informatic Technologies of Russian Acad. Sci., 142092 Troitsk, Moscow Region, Russian Federation

110   Amat A, Rigau J, Nicolaua R, Aalders M et al. Effect of red and near-infrared laser light on adenosine triphosphate (ATP) in the luciferine–luciferase reaction. Journal of Photochemistry and Photobiology A: Chemistry. 2004; 168 (1-2): 59-65.

111   Photomedicine and Laser Surgery;   Ga-As (808 nm) Laser Irradiation Enhances ATP Production in Human Neuronal Cells in Culture; Jun 2007, Vol. 25, No. 3 : 180 -182; U. Oron, Ph.D

112   Efficacy of low level laser therapy on neurosensory recovery after injury to the inferior alveolar nerve. Tuncer Ozen, Kaan Orhan,corresponding author. Ilker Gorur, Adnan Ozturk. Head Face Med. 2006; 2: 3.

113   Promotion of regenerative processes in injured peripheral nerve induced by low-level laser therapy. Photomed Laser Surg. 2007 Apr;25(2):107-11. Mohammed IF, Al-Mustawfi N, Kaka LN. Department of Anatomy, Al-Kindy Medical College, Baghdad University, Baghdad, Iraq.

114   Effect of laser radiation on the posttraumatic recovery of the interneuronal connections in the sympathetic trunk;Rakhishev AR, Usupbekova BSh. Arkh Anat Gistol Embriol. 1987 Sep;93(9):58-63.

115   Effect of low-level laser treatment on neurosensory deficits subsequent to saggittal split ramus osteotomy; Shelley M. Khullar, Babak Emami, Anders Westermark, Hans R. Haans Oral Surgery, Oral Medicine, Oral Pathology, Oral Radiology and Endodontics. August 1996 (Vol. 82, Issue 2, Pages 132-138).

116   Topical hyperbaric oxygen and low energy laser therapy for chronic diabetic foot ulcers resistant to conventional treatment.Landau Z, Schattner A. Yale J Biol Med. 2001 Mar-Apr;74(2):95-100.

117   Diabetic neuropathic foot ulcer: successful treatment by low-intensity laser therapy. Schindl A, Schindl M, Pernerstorfer-Schön H, Kerschan K, Knobler R, Schindl L. Dermatology. 1999;198(3):314-6.

118   Low-intensity laser irradiation improves skin circulation in patients with diabetic microangiopathy. Schindl A, Schindl M, Schön H, Knobler R, Havelec L, Schindl L. Diabetes Care. 1998 Apr;21(4):580-4.

119   The diabetic foot. General considerations and proposal of a new therapeutic and preventive approach. Rinaldi F, Alberetto M, Pontiroli A. Diabetes Res Clin Pract. 1993 Jul;21(1):43-9.

120   The use of helium-neon laser radiation in the treatment of trophic disorders in patients with diabetes mellitus; Zubkova ST. Klin Khir. 1992;(3):47-9.

121   Intravascular laser irradiation of blood in complex treatment of obliterating atherosclerosis of the lower extremity vessels in elderly and senile patients. Iaitskiĭ NA, Ageenko EM, Davydenko TE, Volchkov VA, Churzin OA, Zharskaia VD.Vestn Khir Im I I Grek. 2006;165(4):34-7.

122   Possibilities of the low-intensity infrared laser therapy in atherosclerotic lesions of the cardiovascular system; Tsyb AF, Kaplan MA, Tkachenko NP, Popovkina OE. Vestn Ross Akad Med Nauk. 2003;(12):45-52.

123   Low-intensive laser irradiation in combined treatment of lower limbs atherosclerotic lesions. HHLipatova IO, Arslanova VM, Kriuchkov VI, Markov AN, Sakharov AB.HH Khirurgiia (Mosk). 2003;(4):14-9.

124   Low intensity laser radiation in complex therapy of patients with vascular obliterating atherosclerosis of low extremities HHKlimenko IT, Shuvalova IN.HH Lik Sprava. 2002;(8):98-102.

125   The micro circulation of patients with arterial ischemia of the lower extremities during laser therapy; Kozlov VI, Soboleva TM, Azizov GA, Len'kova NA, Elfimov AI, Iskakova ZhT.; Fiziol Zh SSSR Im I M Sechenova. 1991 Jun;77(6):55-67. The laser therapy proved to be successful in the patients with the II and III stages of arterial ischemia.

126   GaAlAs (830 nm) low-level laser enhances peripheral endogenous opioid analgesia in rats. Hagiwara S, Iwasaka H, Okuda K, Noguchi T.; Lasers Surg Med. 2007 Dec;39(10):797-802.

127   Analgesic effect of He-Ne (632.8 nm) low-level laser therapy on acute inflammatory pain. Ferreira DM, Zângaro RA, Villaverde AB, Cury Y, Frigo L, Picolo G, Longo I, Barbosa DG. Photomed Laser Surg. 2005 Apr;23(2):177-81.

128   A systematic review of common conservative therapies for arm lymphoedema secondary to breast cancer treatment. Moseley AL, Carati CJ, Piller NB. Ann Oncol. 2007 Apr;18(4):639-46. Epub 2006 Oct 3.

129   Effects of low power laser and low dose amitriptyline therapy on clinical symptoms and quality of life in fibromyalgia: a single-blind, placebo-controlled trial. Gür A, Karakoc M, Nas K, Cevik R, Sarac J, Ataoglu S. Rheumatol Int. 2002 Sep;22(5):188-93. Epub 2002 Jul 6.

130   Efficacy of low power laser therapy in fibromyalgia: a single-blind, placebo-controlled trial. Gür A, Karakoç M, Nas K, Cevik R, Saraç J, Demir E. Lasers Med Sci. 2002;17(1):57-61.

131   Effects of helium-neon laser irradiation on skin resistance and pain in patients with trigger points in the neck or back. Snyder-Mackler L, Barry AJ, Perkins AI, Soucek MD. Phys Ther. 1989 May;69(5):336-41.

132   Diode laser in cervical myofascial pain: a double-blind study versus placebo. Ceccherelli F, Altafini L, Lo Castro G, Avila A, Ambrosio F, Giron GP. Clin J Pain. 1989 Dec;5(4):301-4.

133   Efficacy of 904 nm gallium arsenide low level laser therapy in the management of chronic myofascial pain in the neck: a double-blind and randomize-controlled trial. Gur A, Sarac AJ, Cevik R, Altindag O, Sarac S. Lasers Surg Med. 2004;35(3):229-35.

134   Effect of low level laser therapy in rheumatoid arthritis patients with carpal tunnel syndrome. Ekim A, Armagan O, Tascioglu F, Oner C, Colak M.Swiss Med Wkly. 2007 Jun 16;137(23-24):347-52.

135   Experimental photodynamic laser therapy for rheumatoid arthritis with a second generation photosensitizer. Hendrich C, Huttmann G, Vispo-Seara JL, Houserek S, Siebert WE. Knee Surg Sports Traumatol Arthrosc. 2000;8(3):190-4.

136   Laser therapy of rheumatoid arthritis. Goldman JA, Chiapella J, Casey H, Bass N, Graham J, McClatchey W, Dronavalli RV, Brown R, Bennett WJ, Miller SB, Wilson CH, Pearson B, Haun C, Persinski L, Huey H, Muckerheide M. Lasers Surg Med. 1980;1(1):93–101.

137   Soft-laser therapy of rheumatoid arthritis. Bliddal H, Hellesen C, Ditlevsen P, Asselberghs J, Lyager L. Scand J Rheumatol. 1987;16 (4):225–228.

138   A double blind randomised trial of low power laser treatment in rheumatoid arthritis. Ann Rheum Dis. 1993 October; 52(10): 703–706. J K Heussler, G Hinchey, E Margiotta, R Quinn, P Butler, J Martin, and A D Sturgess Dept of Rheumatology, St George Hospital, Sydney, Australia. This is one of the silliest things I have seen. The researchers threw this out because although 72% of the patients reported significant pain relief, the untreated hand got better as well, so they thought it was "all in their heads", (my quotes.) Didn't they read the literature that shows that in rheumatoid arthritis we get a systemic effect?) It's almost as if someone *wanted* it to be negative.

139   Clinical application of GaAlAs 830 nm diode laser in treatment of rheumatoid arthritis, Laser Therapy 3: 77-82, 1991. Moore.

140   Low level laser therapy in shoulder tendinitis/bursitis, epicondylalgia and ankle sprain. Jan M Bjordal in 1997, at the Division of Physiotherapy Science, University of Bergen. It has also been published in Physical Therapy; Reviews. 1998; 3: 121-132.

141   Effectiveness of rehabilitation for patients with subacromial impingement syndrome: a systematic review. Lori A Michener, Matthew K Walsworth, Evie N Burnet J Hand Ther (Vol. 17, Issue 2, Pages 152-64)

142   A randomised, placebo controlled trial of low level laser therapy for activated Achilles tendinitis with microdialysis measurement of peritendinous prostaglandin E2 concentrations. Bjordal JM, Lopes-Martins RA, Iversen VV. Br J Sports Med. 2006 Jan;40(1):76-80; discussion 76-80.

143   Low-level laser therapy (LLLT) prevents oxidative stress and reduces fibrosis in rat traumatized Achilles tendon. Fillipin LI, Mauriz JL, Vedovelli K, Moreira AJ, Zettler CG, Lech O, Marroni NP, Gonzalez-Gallego J Lasers Surg Med. 2005 Oct;37(4):293-300.

144   Low-level laser therapy improves early healing of medial collateral ligament injuries in rats. Photomed Laser Surg. 2005 Dec;23(6):556-60. Bayat M, Delbari A, Almaseyeh MA, Sadeghi Y, Bayat M, Reziae F.

145   Low power laser therapy of shoulder tendinitis. S England, A J Farrell, J S Coppock, G Struthers, P A Bacon Scand J Rheumatol; 1989 (Vol. 18, Issue 6, Pages 427-31)

146   Efficacy of low-intensity laser radiation and antibacterial therapy in the treatment of chronic prostatitis in the presence of sexually transmitted infections; Neĭmark AI, Khrianin AA, Safina ON, Neĭmark BA, Kondrat'eva IuS. Urologiia. 2007 May-Jun;(3):50-4, 56.

147   A method and the clinico-laboratory results of laser therapy for patients with chronic prostatitis; Shabad AL, Red'kovich VI, Safarov RM. Urol Nefrol (Mosk). 1994 Nov-Dec;(6):26-9.

148   Using AELTIS-synchro-02 device in the therapy of chronic bacterial prostatitis. Orlov VN, Kozboda AS, Kravchenko VV, Kalinina SA. Urologiia. 2006 Jul-Aug;(4):54-7.

149   Electrolaser therapy on the Iarilo device in patients with chronic chlamydial prostatitis. Tiktinskiĭ OL, Kalinina SN, Novikova LI, Mishanin EA, Tiktinskiĭ NO. Urol Nefrol (Mosk). 1997 Jul-Aug;(4):25-9.

150   Treatment with neodymium: YAG laser in patients with chronic prostatitis: a preliminary report. Serel TA, Koşar A, Ozturk A, Dogruer K, Tahoğlu M, Kecelioğlu M. Int Urol Nephrol. 1997;29(1):53-8.

151   Low level laser irradiation (HE-Ne) improved functional preservation of the isolated rat heart. Journal of Electrocardiology, Volume 40, Issue 4, Pages S4-S5   N. Takzaree, A. Takzaree, N. Daneshvar, R. Takzaree

152   Dynamics of hyperlipidemia and peripheral blood flow in patients with diabetes mellitus after the course of combined laser therapy in ambulatory-polyclinic conditions. Oprysko T V et al. Laser & Health '99, December 8-10 1999, Moscow. 218 patients with DM were treated with laser blood irradiation. 93 patients had DM I type and 125 DM II type. A HeNe laser of 2 mW was used intravenously. In addition an infrared 890 nm laser (5-20 mW) was used for irradiation over the projections of the liver, spleen and pancreas. Treatment was given daily for a period of 8 days. Repeated sessions were given at 3 and 6 months. Sugar correcting medications were decreased 200%. From the first day the patients' extremities grew warmer, pain decreased and symptoms of encephalopathy decreased. Levels of total blood cholesterol, LDL cholesterol and triglycerides decreased to normal values with a simultaneous increase of alhpa-LP. Sugar concentration in blood also decreased.

153   Comparison between Wound Healing in Induced Diabetic and Non diabetic Rats after Low-Level Laser Therapy Aug 2006, Vol. 24, No. 4: 474-479   Sylvia Bicalho Rabelo , D.D.S., M.S.D. Instituto de Pesquisa e Desenvolvimento (IP&D), Universidade do Vale do Paraiba

154   Systemic Effects of Low-Intensity Laser Irradiation on Skin Micro circulation in Patients with Diabetic MicroangiopathyAndreas Schindl, Georg Heinze, Martin Schindl, Heidemarie Pernerstorfer-Schönd and Liesbeth Schindl Microvascular Research Volume 64, Issue 2, September 2002, Pages 240-246

155   Effects of near-infrared low-level laser irradiation on micro circulation Yasuyo Maegawa, MD, Toshiyuki Itoh, MD, PhD, Toyoshi Hosokawa, MD, PhD, Kazuhiro Yaegashi, MD, PhD, Mayumi Nishi, MD, PhD. Lasers in Surgery and Medicine Volume 27 Issue 5, Pages 427 - 437

156 Low Level Laser Therapy in the Treatment of Carpal Tunnel Syndrome. Thomas E. Anderson, PhD Automotive Safety and Health Research NAO Research and Development Center. Good WT, Kerr HH, Shumaker B, Bendick PJ, Nolta RG. 25-Jan-95

157 Photobiomodulation of pain in carpal tunnel syndrome: review of seven laser therapy studies. Naeser MA.Photomed Laser Surg. 2006 Apr;24(2):101-10.

158 Effect of low level laser therapy in rheumatoid arthritis patients with carpal tunnel syndrome. Ekim A, Armagan O, Tascioglu F, Oner C, Colak M. Swiss Med Wkly. 2007 Jun 16;137(23-24):347-52.

159 Treatment of carpal tunnel syndrome by low-level laser versus open carpal tunnel release. Lasers Med Sci. 2007 Nov;22(4):265-70. Epub 2007 Mar 3. Elwakil TF, Elazzazi A, Shokeir H. Unit of General Surgery, National Institute of Laser Enhanced Sciences (NILES), Cairo University, Egypt.

160 Exocrine and endocrine testicular function during the treatment of experimental orchitis and non-specific orchoepididymitis by low-energy laser radiation. Leonid L. Reznikov Proceedings of SPIE -- Volume 2391; Laser-Tissue Interaction VI, Steven L. Jacques, Editor, May 1995, pp. 641-647; Univ. of Colorado Health Sciences Ctr. (USA) Ludmila S. Pupkova St. Petersburg Scientific Research Ctr. (Russia) H. Bell Embryology Associates, Inc. (USA) Alexander G. Murzin St. Petersburg Scientific Research Ctr. (Russia)

161 Quantitative and qualitative changes of the seminiferous epithelium induced by Ga. Al. As. (830 nm) laser radiation. Taha MF, Valojerdi MR. Lasers Surg Med. 2004;34(4):352-9.

162 Efficacy of low-intensity laser radiation and antibacterial therapy in the treatment of chronic prostatitis; in the presence of sexually transmitted infections. Neĭmark AI, Khrianin AA, Safina ON, Neĭmark BA, Kondrat'eva IuS. Urologiia. 2007 May-Jun;(3):50-4, 56.

163 A method and the clinico-laboratory results of laser therapy for patients with chronic prostatitis; Shabad AL, Red'kovich VI, Safarov RM. Urol Nefrol (Mosk). 1994 Nov-Dec;(6):26-9.

164 Using AELTIS-synchro-02 device in the therapy of chronic bacterial prostatitis;Orlov VN, Kozboda AS, Kravchenko VV, Kalinina SA. Urologiia. 2006 Jul-Aug;(4):54-7.

165 Electro-laser therapy on the Iarilo device in patients with chronic chlamydial prostatitis; Tiktinskiĭ OL, Kalinina SN, Novikova LI, Mishanin EA, Tiktinskiĭ NO. Urol Nefrol (Mosk). 1997 Jul-Aug;(4):25-9.

166 Treatment with neodymium: YAG laser in patients with chronic prostatitis: a preliminary report. Serel TA, Koşar A, Ozturk A, Dogruer K, Tahoğlu M, Kecelioğlu M. Int Urol Nephrol 1997;29(1):53-8.

167 Semiconductor laser rays therapy for the treatment of chronic prostatitis. Prof. Guido Strada, Dr med. Franco Gadda, Dr med. Paolo Dell'Orto, Dr med. Marilena Casu, Dr med. Andrea Baccalin Department of Urology, San Gerardo Hospital, Monza (Milan-Italy) The Journal of Laser Therapy Millennium Edition Volumes 12 & 13

168 Effcets of GaAs-lasers combined with radiotherapy on murine sarcoma deoends on tumor size. Saldo I et al. Lasers in Surgery and Medicine. 1989; Suppl 1:40.

169 Preoperative activation of the immune systems by low recative level laser therapy (laser theyapy) in oncologic patient: A preliminary report. Skobelkin OK et al. Laser Therapy. 1991;3(4):169-176

170 Noninvasive laser therapy of morbus peyronie – induratio penis plastica. Miroslav Prochazka, M. D., Karel Koci, M. D. Rehabilitation Clinic Jarov, Prague; Andrology Clinic Andromeda, Prague, Czech Republic

171 The testosterone connection. S Smith. Kansas Nurse, Aug 2003

172 Proceedings of SPIE -- Volume 2391; Laser-Tissue Interaction VI, Steven L. Jacques, Editor, May 1995, pp. 641-647; Exocrine and endocrine testicular function during the treatment of experimental orchitis and nonspecific orchoepididymitis by low-energy laser radiation.

173 Effects of low-power laser radiation on mice immunity. E G Novoselova, O V Glushkova, D A Cherenkov, V M Chudnovsky, E E Fesenko. Photodermatol Photoimmunol Photomed Feb 2006 (Vol. 22, Issue 1, Pages 33-8)

174 Zinc Levels and Blood Enzyme Activities in Egyptian Male Subjects with Retarded Growth and Sexual Development . YD Coble, R Vanreen, AR Schulert, RP Koshakji, Z Farad, JT Davis. American Journal of Clinical Nutrition, Vol 19, 415-421

175 Chronic inflammation and cancer. Oncology (Williston Park). 2002 Feb;16(2):217-26, 229; discussion 230-2. Shacter E, Weitzman SA.Laboratory of Immunology, Center for Biologics Evaluation and Research, Food and Drug Administration, Bethesda, Maryland 20892-4555, USA.

176   Inflammation and cancer: an epidemiological perspective.Thun MJ, Henley SJ, Gansler T. Novartis Found Symp. 2004;256:6-21; discussion 22-8, 49-52, 266-9. Epidemiology and Surveillance Research, American Cancer Society, 1599 Clifton Road, Atlanta, GA 30329-4251, USA.

177   DNA damage induced by chronic inflammation contributes to colon carcinogenesis in mice Lisiane B. Meira, James M. Bugni, Stephanie L. Green, Chung-Wei Lee, Bo Pang, Diana Borenshtein, Barry H. Rickman, Arlin B. Rogers, Catherine A. Moroski-Erkul, Jose L. McFaline, David B. Schauer, Peter C. Dedon, James G. Fox and Leona D. Samson J. Clin. Invest. doi:10.1172/JCI35073.

178   The inflammatory chemokines CCL2 and CCL5 in breast cancer. Soria G, Ben-Baruch A. Cancer Lett. 2008 Apr 23. Department of Cell Research and Immunology, George S. Wise Faculty of Life Sciences, Tel Aviv University, Tel Aviv 69978, Israel.

179   Low-Intensity Laser Therapy for Benign Fibrotic Lumps in the Breast Following Reduction Mammaplasty EL Nussbaum, PT, is Assistant Professor, Department of Physical Therapy, University of Toronto, and Academic Practice Leader, Mount Sinai Hospital, Toronto, Canada. PHYS THER Vol. 79, No. 7, July 1999, pp. 691-698

180   A systematic review of common conservative therapies for arm lymphoedema secondary to breast cancer treatment. Moseley AL, Carati CJ, Piller NB.   Ann Oncol. 2007 Apr;18(4):639-46. Epub 2006 Oct 3.

181   FDA Clears Laser-Based Lymphedema Therapy   Oct/Nov 2006 News Release; The Food and Drug Administration (FDA) has cleared the RianCorp LTU-904 low-level laser device as an additional weapon in the battle for better quality of life for people living with lymphedema. After a four-year evaluation, the FDA cleared the LTU-904 for use as part of a therapy regime to treat post-mastectomy lymphedema. FDA clearance permits therapists throughout the USA to use the handheld battery-powered device to treat a patient with lymphedema

182   Magnetic resonance imaging (MRI) controlled outcome of side effects caused by ionizing radiation, treated with 780 nm-diode laser - Preliminary results. Journal of Photochemistry and Photobiology B: Biology. 2000; 59/1-3 (1-8) Schaffer-M, Bonel-H, Sroka-R.

183   Chiropractic Approach to Premenstrual Syndrome (PMS). Wittler, M. The J Chiro Res & Clin Invest 1992; 8 (2):   26–29

184   Effcets of GaAs-lasers combined with radiotherapy on murine sarcoma deoends on tumor size. Saldo I et al. Lasers in Surgery and Medicine. 1989; Suppl 1:40.

185   Preoperative activation of the immune systems by low recative level laser therapy (laser theyapy) in oncologic patient: A preliminary report. Skobelkin OK et al. Laser Theraoy. 1991;3(4):169-176

186   Oncological aspects of the biological action of low level laser radiation. II. Investigation of primary mechanisms and clinical applications Ivanov A.V., Zakharov S.D. SPIE proceedings series Laser use in oncology II (selected papers on laser use in oncology 1997-1999) 2000, vol. 4059, pp. 101-111, VIII, 230 p.,] (28 ref.) ISBN 0-8194-3688-7

187   Effects of Low-Level Laser Therapy on Malignant Cells: In Vitro Study. Antonio Luiz Barbosa Pinheiro, Silene Carneiro Nascimento, Alessandro Leonardo de Barros Vieira, Aldo Brugnera, Fatima A. Zanin, Aluizio Barros Rolim, Pedro Soriano da Silva. Journal of Clinical Laser Medicine & Surgery. February 1, 2002, 20(1): 23-26.

188   Effects of sarin on the nervous system in rescue team staff members and police officers 3 years after the Tokyo subway sarin attack. Nishiwaki, Y., Maekawa, K., Ogawa, Y., Asukai, N., inami, M.O., & Omae, K. Environmental Health Perspective 109, 1169-1173 (2001).

189   Psycho-physiological effects of the terrorist sarin attack on the Tokyo subway system. Kawana, N. Military Medicine 166, 23-26 (2001).

190   Response of rats to low levels of sarin. Henderson, R.F., Barr, E.B., Blackwell, W.B., Clark, C.R., Conn, C.A., Kalra, R., March, T.H., Sopori, M. L., Tesfaigzi, Y., Menache, M.G., & Mash, D.C. Toxicol.Appl.Pharmacol. 184, 67-76(2002).

191   Gulf war syndrome There may be no specific syndrome, but troops suffer after most wars. F Murphy BMJ 1999 January 30; 318(7179): 274-275

192   Treatment of fibromyalgia with antidepressants A meta-analysis. O'Malley PG, Balden E, Tomkins G, Santoro J, Kroenke K, Jackson JL, Gen Intern Med 2000;15(9):659-66.

193   Common shampoo ingredients found to harm memory cells M Adams August 07, 2006 by: NaturalNews

194   Photoengineering of Tissue Repair in Skeletal and Cardiac Muscles Dr. Uri Oron, Ph.D. Apr 2006, Vol. 24, No. 2: 111-120 , Photomedicine and Laser Surgery. The George S. Wise Faculty of Life Sciences, Tel-Aviv University, Tel-Aviv, Israel.

195 Evaluation Of Quantum Therapy For The Treatment Of Stable Angina   Likarska sprava. 2001; (5-6): 111-114.

196 Influence Of Low Dose Laser Therapy On Endothelial Function In Patients With CAD. S. Belousov, E. Galperin, E. Smetova Vopr Kurortol Fizioter Lech Fiz Kult. 2003 Jul-Aug;(4):10-3.

197 Efficacy Of Laser Therapy In Patients With Ischemic Heart Disease. Klin Med (Mosk). 2002;80(4):31-3.   Vopr Kurortol Fizioter Lech Fiz Kult. 2003 May-Jun;(3):22-5.

198 Antioxidant Action And Therapeutic Efficacy Of Laser Irradiation Of Blood In Patients With Ischemic Heart Disease   Volotovskaia AV, Ulashchik VS, Filipovich VN. Vopr Kurortol Fizioter Lech Fiz Kult. 2002 Jul-Aug;(4):9-11.

199 Effects of laser therapy on ventricular and supraventricular extrasystoles in patients with angina pectoris of 1-3 functional classes. V.A. Lyusov, K.G. Khutchumova, Y.N. Fedullaev The Russian Medical State University, Moscow, Russia

200 The Effectiveness And Influence On Rheological Blood Features And Endothelian Vessels Function Of Infrared Laser Therapy In Patients With Angina Pectoris T.M. Zenkovskaya, O.A. Matronchick, F.A. Khairullina, A.D. Kats Hospital of War Invalids, Perm State Medical Academy, Russia

201 Intravenous A Laser Irradiation Of Blood In Complex Rehabilitation Of The Patients With Ischemic Heart Disease In The Senior Age Groups Y.M. Kazakov, L.A. Zvyaginceva, et al. Ukrainian medical stomatological academy, Poltava, Ukraine

202 The Application Of Low-Intensity Laser Radiation For The Treatment Of Infarctional Patients With Early Angina Pectoris Refractering To The Drug Therapy. Y.L.Gabinsky, Y.R. Yakovlev, S.V. Yakovleva Yekaterinburg Infarction Centre, Yekaterinburg, Russia

203 Efficiency Of Intravascular Blood Irradiation With He-Ne Laser In Patients With Stable Effort Angina Pectoris Depending On Functional Class A.I. Koryakov Sverdlovsk Regional Clinical Psychoneurological Hospital of the Disabled Soldiers, Yekaterinburg, Russia

204 Augmentation of the expression of proangiogenic genes in cardiomyocytes with low dose laser irradiation In vitro Ashwani Khanna, Latha Raja Shankar, Michael H. Keelan, Ran Kornowski, Martin Leon, Jeffrey Moses, Nicholas Kipshidze Cardiovascular Radiation Medicine   July 1999 (Vol. 1, Issue 3, Pages 265-269)

205 Results Of 10-Year Use Of Low Intensity Laser Therapy And Conventional Treatment Of Patients With Stenocardia G.W. Babushkina, I.M. Korochkin, A.V. Kartelishev Bashkir Medical University, Ufa, Russian State Medical University, Moscow, RF

206 Diagnostic Implications Of Changed Red Cell Count In Low-Intensity Laser Radiation Of Blood In Elderly Patients With Coronary Heart Disease Simonenko VB, Siuch NI, Vokuev IA. Vopr Kurortol Fizioter Lech Fiz Kult. 2001 Nov-Dec;(6):10-3.

207 Antioxidant Action And Therapeutic Efficacy Of Laser Irradiation Of Blood In Patients With Ischemic Heart Disease   Volotovskaia AV, Ulashchik VS, Filipovich VN. Vopr Kurortol Fizioter Lech Fiz Kult. 2002 Jul-Aug;(4):9-11.

208 Influence Of Intravenous Blood Irradiation With He-Ne Laser On The Antianginal Effect Of Nitroglycerin In Patients With Stable Effort Angina Pectoris. A.I. Koryakov, N.N. Koryakova, A.I. Iofin Vopr Kurortol Fizioter Lech Fiz Kult. 2002 Jul-Aug;(4):9-11.

209 The micro circulation of patients with arterial ischemia of the lower extremities during laser therapy Kozlov VI, Soboleva TM, Azizov GA, Len'kova NA, Elfimov AI, Iskakova ZhT. Fiziol Zh SSSR Im I M Sechenova. 1991 Jun;77(6):55-67.

210 Membranotropic Effect of Low-Intensity Laser Radiation of the Blood. A. A. Spasov, V. V. Nedogoda, O. V. Ostrovskii, Kuame Konan. Bulletin of Experimental Biology and Medicine. Vol. 126, No. 10, pp. 412–415, October, 1998

211 Modification in relationships in the system endotherlial cell-blood platelet by low level laser irradiation. G.E.Brill , L.V.Gasparyan, A.M.Makela. Laser Florence 2003

212 Dynamics of lipid metabolism and peripheral blood flow rates in patients with atherosclerosis in conjunction with renal dysfunction after the course of combined laser therapy. Kovalyova T V et al. Urol Nefrol (Mosk). 1997 Mar-Apr;(2):12-3.

213 Laser correction of micro circulation disorders in patients having CHD with hypercholesterinemia A P Vasil'ev , M A Sekisova , N N Strel'tsova , Iu N Senatorov

214 Laser blood irradiation effect on electrophysiological characteristics of acute coronary syndrome patients. Khotiaintsev, Konstantin S.; Doger-Guerrero, E.; Glebova, L.; Svirid, V.; Sirenko, Yuri Proc. SPIE Vol. 2929, p. 132-137, Effects of Low-Power Light on Biological Systems II, Giulio Jori; Tina I. Karu; Eds. 11/1996

132

215   In vitro effects of helium-neon laser irradiation on human blood: blood viscosity and deformability of erythrocytes.Mi XQ, Chen JY, Liang ZJ, Zhou LW. - Department of Physics, State Key Laboratory of Applied Surface Physics, Fudan University, Shanghai, China.

216   Low level laser treatment of primary and secondary Raynaud's phenomenon. Al-Awami M, Schillinger M, Gschwandtner M.E, Maca T, Haumer M, minare E. - Department of Medical Angiology, Univ of Vienna, Austria

217   Low power laser protects human erythrocytes In an In vitro model of artificial heart-lung machines. Artificial Organs. 2000; 24 (11): 870-3. Itoh-T, Murakami-H, Orihashi-K et al.

218   Helium-neon and nitrogen laser irradiation accelerates the phagocytic activity of human monocytes. Hemvani N, Chitnis DS, Bhagwanani NS. - Dept of Micro-biology and Immunology, Choithram Hospital & Research Centre, Indore, India.

219   Role of the circulating blood in initiation of therapeutic effects of visible light.K.A.Samoilova Institute of Cytology of the Russian Academy of Sciences, St. Petersburg, Russia The Journal of Laser Therapy. Millennium Edition Volumes 12 & 13 .

220   Effects of low-power laser radiation on mice immunity. E G Novoselova, O V Glushkova, D A Cherenkov, V M Chudnovsky, E E Fesenko. Photodermatol Photoimmunol Photomed Feb 2006 (Vol. 22, Issue 1, Pages 33-8)

221   Oncological aspects of the biological action of low level laser radiation. II. Investigation of primary mechanisms and clinical applications IVANOV A. V. ; ZAKHAROV S. D. SPIE proceedings series Laser use in oncology II : (selected papers on laser use in oncology 1997-1999) 2000, vol. 4059, pp. 101-111, [Note(s) : VIII, 230 p.,] (28 ref.) ISBN 0-8194-3688-7

222   Transcranial application of low-energy laser irradiation improves neurological deficits in rats following acute stroke. Luis Detaboada, Sanja Ilic, Sandra Leichliter-Martha, Uri Oron, Amir Oron, Jackson Streeter Lasers Surg Med Jan 2006 (Vol. 38, Issue 1, Pages 70-3)

223   Low-level laser therapy applied transcranially to rats after induction of stroke significantly reduces long-term neurological deficits. Amir Oron, Uri Oron, Jieli Chen, Anda Eilam, Chunling Zhang, Menachem Sadeh, Yair Lampl, Jackson Streeter, Luis DeTaboada, Michael Chopp. Stroke Oct 2006 (Vol. 37, Issue 10, Pages 2620-4)

224   Transcranial infrared laser therapy improves clinical rating scores after embolic strokes in rabbits. Paul A Lapchak, Jiandong Wei, Justin A Zivin Stroke Aug 2004 (Vol. 35, Issue 8, Pages 1985-8)

225   Laser treatment for stroke.Lampl Y. Expert Rev Neurother. 2007 Aug;7(8):961-5.

226   Treatment of experimentally induced transient cerebral ischemia with low energy laser inhibits nitric oxide synthase activity and up-regulates the expression of transforming growth factor-beta 1. Mason C P Leung, Samuel C L Lo, Flora K W Siu, K-F So Lasers Surg Med   2002 (Vol. 31, Issue 4, Pages 283-8)

227   Phospholipid pool, lipid peroxidation, and superoxide dismutase activity under various types of oxidative stress of the brain and the effect of low-energy infrared laser irradiation. Karageuzyan KG, Sekoyan ES, Karagyan AT, Pogosyan NR, Manucharyan GG, Sekoyan AE, Tunyan AY, Boyajyan VG, Karageuzyan MK Biochemistry (Mosc) 1998; 63:1226-32.

228   Infrared laser therapy for ischemic stroke: a new treatment strategy: results of the NeuroThera Effectiveness and Safety Trial-1 (NEST-1).Lampl Y, Zivin JA, Fisher M, Lew R, Welin L, Dahlof B, Borenstein P, Andersson B, Perez J, Caparo C, Ilic S, Oron U Stroke Jun 2007 (Vol. 38, Issue 6, Pages 1843-9)

229   Effects of power densities, continuous and pulse frequencies, and number of sessions of low-level laser therapy on intact rat brain.Ilic S, Leichliter S, Streeter J, Oron A, DeTaboada L, Oron U Photomed Laser Surg 2006; 24:458-66.

230   Low-level laser therapy applied transcranially to mice following traumatic brain injury significantly reduces long-term neurological deficits. Oron A, Oron U, Streeter J, de Taboada L, Alexandrovich A, Trembovler V, Shohami E J Neurotrauma 2007; 24:651-6.

231   Quantitative and qualitative changes of the seminiferous epithelium induced by Ga. Al. As. (830 nm) laser radiation. Taha MF, Valojerdi MR.Lasers Surg Med. 2004;34(4):352-9.

232   Evaluation of low intensity laser effects on the thyroid gland of male mice. Luciane Hiramatsu Azevedo, Ana Cecília Correa Aranha, Sandro Fernando Stolf, Carlos de Paula Eduardo, Martha Marques Ferreira Vieira Photomed Laser Surg Dec 2005 (Vol. 23, Issue 6, Pages 567-70)

233   The effect of low-intensity laser radiation in the infrared and red ranges on arterial pressure regulation in patients with borderline hypertension Shuvalova IN, Klimenko IT, Zhukova LP, Oborin IuI.

133

234 Efficiency of low-intensity laser radiation in essential hypertension Velizhanina IA, Gapon LI, Shabalina MS, Kamalova NN

235 The effect of low-intensity laser radiation in the infrared and red ranges on arterial pressure regulation in patients with borderline hypertension Shuvalova IN, Klimenko IT, Zhukova LP, Oborin IuI Lik Sprava ; :141-3.

236 Registration of surface fluorescence of endogenous porphyrins during low-level laser therapy of gastric and duodenal ulcers. G I Klebanov, D A Rogatkin, S G Tereshchenko Biofizika. Vol. 49, Issue 5, Pages 941-7

237 Clinical and immunological aspects of low-intensity laser irradiation in patients with gastroduodenal ulcers A P Alebastrov, M A Butov, P A Chumachenko Eksp Klin Gastroenterol 2004 (Issue 4, Pages 33-108)

238 Potentialities of alternative non-drug therapy of gastric ulcer A P Alebastrov, M A Butov Klin Med (Mosk) 2005 (Vol. 83, Issue 11, Pages 69-71)

239 Effects of He-Ne laser irradiation on chronic atrophic gastritis in rats. Xue-Hui Shao, Yue-Ping Yang, Jie Dai, Jing-Fang Wu, Ai-Hua Bo World J Gastroenterol Jul 2005 (Vol. 11, Issue 25, Pages 3958-61)

240 Correction of the gastric and duodenal motility by laser puncture in pancreatobiliary diseases. A A Gontar Lik Sprava 2002 (Issue 1, Pages 126-9) Used low level laser on acupuncture points.

241 Effect of impulse infrared laser radiation on bioenergetic metabolism in gastric mucosa in patients with gastroduodenal ulcer. F I Khamrabaeva, A A Khodzhimatov Vopr Kurortol Fizioter Lech Fiz Kult (Issue 5, Pages 26-7)

242 Biospectroscopy-based endoscopy with low-intensive laser radiation for differential diagnostics and treatment of peptic ulcer. S G Tereshchenko, D A Rogatkin, L G Lapaeva, P V Astakhov Vestn Ross Akad Med Nauk 2005 (Issue 5, Pages 28-31)

243 Effect of low-energy laser radiation on the levels of cyclic nucleotides and lactic acid of the mucous membranes in gastric and duodenal ulcers. A S Balalykin, T V Terekhova, S V Lapshina Klin Med (Mosk) Mar 1989 (Vol. 67, Issue 3, Pages 108-10)

244 Possibilities of the treatment of biliary dyskinesia by laser irradiation. L P Vorob'ev, L M Salova, V M Meshkov, G V Chubarov Klin Med (Mosk) May 1990 (Vol. 68, Issue 5, Pages 110-5)

245 Effect of the clinical application of the GaAlAs laser in the treatment of dentine hypersensitivity. Ana Lucia Marsilio, José Roberto Rodrigues, Alessandra Bühler Borges J Clin Laser Med Surg Oct 2003 (Vol. 21, Issue 5, Pages 291-6)

246 The effectiveness of the Nd:YAG laser in the treatment of dental hypersensitivity.Gelskey SC, White JM, Pruthi VK.J Can Dent Assoc. 1993 Apr;59(4):377-8, 383-6.

247 A clinical investigation of the efficacy of low level laser therapy in reducing orthodontic postadjustment pain. H M Lim, K K Lew, D K Tay Am J Orthod Dentofacial Orthop Dec 1995 (Vol. 108, Issue 6, Pages 614-22)

248 Effects of low-power red laser on dentine–pulp interface after cavity preparation. An ultrastructural study Bruno Miranda Godoy, Victor Elias Arana-Chavez, Silvia Cristina Núñez, Martha Simões Ribeiro Archives of Oral Biology September 2007 (Vol. 52, Issue 9, Pages 899-903)

249 Effect of GaAIAs Laser on Reactional Dentinogenesis Induction in Human Teeth Adriana Nayme Segovia Ferreira, Landulfo Silveira, Walter Joao Genovese, Vera Cavalcante de Araujo, Lucio Frigo, Ricardo Alves de Mesquita, Eduardo Guedes. Photomedicine and Laser Surgery. June 1, 2006, 24(3): 358-365. doi:10.1089/pho.2006.24.358.

250 Efficacy of low level laser therapy in reducing postoperative pain after endodontic surgery—A randomized double blind clinical study. M. B. Kreislera, f1, H. Al Haja, N. Noroozia, B. Willershausenb and B. D'hoedta International Journal of Oral and Maxillofacial Surgery Volume 33, Issue 1, February 2004, Pages 38-41

251 Low-level laser therapy stimulates bone–implant interaction: an experimental study in rabbits Clin Oral Implants Res. 2004; 15 (3): 325-332. Khadra M, Ronold H J, Lyngstadaas S P, Ellingsen J E, Haanaes H R.

252 Osseointegration Of Endosseous Ceramic Implants After Postoperative Low-Power Laser Stimulation: An In Vivo Comparative Study Guzzardella GA, Torricelli P, Nicoli-Aldini N, Giardino R.Department of Experimental Surgery/Codivilla-Putti Research Institute, Rizzoli Orthopaedic Institute, Bologna, Italy. Clin Oral Implants Res. 2003 Apr;14(2):226-32.

134

253 Enhancement of bone formation in rat calvarial bone defects using low-level laser therapy. Oral Surgery, Oral Medicine, Oral Pathology, Oral Radiology, & Endodontics. 97(6):693-700, June 2004. Khadra, Maawan S, MSc a; Kasem, Nesrin S, MSc b; Haanaes, Hans R. DMD, MD, PhD c; Ellingsen, Jan E. S, PhD d; Lyngstadaas, Stale P. S,

254 Effect Of Low-Power Laser Irradiation On Bony Implant Sites. Dortbudak O, Haas R, Mailath-Pokorny G.Department of Oral Surgery, Dental School, University of Vienna, Austria.Clin Oral Implants Res. 2002 Jun;13(3):288-92.

255 Effects Of Visible NIR Low Intensity Laser On Implant Osseointegration In Vivo. Laser Med Surg Abstract issue, 2002: 11. Blay A, Blay C C, Groth E B et al.

256 Influence of low-level laser treatment on bone regeneration and osseointegration of dental implants following sinus augmentation: an experimental study on sheep. Jakse N, Payer M, Tangl S, Berghold A, Kirmeier R, Lorenzoni M. Clin. Oral Impl. Res. 18, 2007; 517-524

257 Effect of Laser Therapy on Bone Tissue Submitted to Radiotherapy: Experimental Study in Rats Jun 2007, Vol. 25, No. 3 : 197 -204 Samantha Seara Da Cunha, D.D.S., Ph.D.

258 Photoengineering of Bone Repair Processes Apr 2006, Vol. 24, No. 2: 169-178, Photomedicine and Laser Surgery Dr. Antonio Luiz B. Pinheiro, D.D.S., Ph.D. Marleny Elizabeth M.M. Gerbi, Ph.D.

259 Effect Of Low-Power GaAlAs Laser (660 Nm) On Bone Structure And Cell Activity: An Experimental Animal Study Nicola RA, Jorgetti V, Rigau J, Pacheco MT, dos Reis LM, Zangaro RA. Lasers Med Sci. 2003;18(2):89-94.

260 Enhancement of bone formation in rat calvarial bone defects using low-level laser therapy. Oral Surgery, Oral Medicine, Oral Pathology, Oral Radiology, & Endodontics. 97(6):693-700, June 2004. Khadra, Maawan S, MSc; Kasem, Nesrin S, MSc; Haanaes, Hans R. DMD, MD, PhD; Ellingsen, Jan E. S, PhD; Lyngstadaas, Stale P.

261 Laser Therapy Plays A Role In Bone Healing. Lasers Surg Med. 1998; 22: 97-102. Luger et al.

262 Computerized Morphometric Assessment Of The Effect Of Low-Level Laser Therapy On Bone Repair: An Experimental Animal Study Silva Júnior AN, Pinheiro AL, Oliveira MG, Weismann R, Ramalho LM, Nicolau RA. J Clin Laser Med Surg. 2002; 20: 83-87

263 Low-Power Laser Irradiation Improves Histomorphometrical Parameters and Bone Matrix Organization During Tibia Wound Healing In Rats Garavello-Freitas I, Baranauskas V, Joazeiro PP, Padovani CR, Dal Pai-Silva M,da Cruz-Hofling MA. J Photochem Photobiol B. 2003 May-Jun;70(2):81-9.

264 Effect Of 830-Nm Laser Light On The Repair Of Bone Defects Grafted With Inorganic Bovine Bone And Decalcified Cortical Osseous Membrane Barbos Pinheiro AL, Limeira Junior Fde A, Marquez Gerbi ME, Pedreira Ramalho LM, Marzola C, Carneiro Ponzi EA, Oliveira Soares A, Bandeira De Carvalho LC, Vieira Lima HC, Oliveira Goncalves T. Laser Center, School of Dentistry, Federal University of Bahia, Brazil. J Clin Laser Med Surg. 2003 Dec;21(6):383-8.

265 Effects Of Pulse Frequency Of Low-Level Laser Therapy (LLLT) On Bone Nodule Formation In Rat Calvarial Cells Department of Orthodontics, Nihon University School of Dentistry at Matsudo Chiba, Japan. J Clin Laser Med Surg. 2003 Oct;21(5):271-7.

266 Laser technology in orthopedics: preliminary study on low power laser therapy to improve the bone-biomaterial interface Guzzardella GA, Torricelli P, Nicoli Aldini N, Giardino R. Int J Artif Organs. 2001 Dec;24(12):898-902.

267 Magnesium deficiency and sudden death
Mark J. Eisenberg, Am Heart J, 102 Aug;124(2):544-9

268 Relations between cardiovascular reactions and glucocorticoid functions of the adrenal cortex as affected by diffuse light of low-intensity helium-neon lasers. Ushkova IN, Pokrovskaia LA, Stepanov GS, Suvorov IM, Kogan ME, Grishina EF. Gig Tr Prof Zabol. 1983 Nov;(11):47-8.

269 Direct action of low-intensity laser irradiation on the morphofunctional status of the zona fasciculata of the adrenal cortex of white rats. V I Kozlov, M K Pugachev, O A Terman Biull Eksp Biol Med Jun 1990 (Vol. 109, Issue 6, Pages 598-600)

270 Prenatal anxiety predicts individual differences in cortisol in pre-adolescent children. Thomas G O'Connor, Yoav Ben-Shlomo, Jon Heron, Jean Golding, Diana Adams, Vivette Glover. Biol Psychiatry Aug 2005 (Vol. 58, Issue 3, Pages 211-7)

271 Exposure to postnatal depression predicts elevated cortisol in adolescent offspring. SL Halligan, Joe Herbert, Ian M Goodyer, Lynne Murray Biol Psychiatry. Feb 2004 (Vol. 55, Issue 4, Pages 376-81)

272   Smoking during pregnancy and newborn neurobehavior. Karen L Law, LR Stroud, LL LaGasse, Raymond Niaura, Jing Liu, Barry M Lester Pediatrics Jun 2003 (Vol. 111, Issue 6 Pt 1, Pages 1318-23)

273   Dysregulation of the Hypothalamic-Pituitary-Adrenal Axis in Postpartum Depression Sandra N. Jolley, PhD, Shawn Elmore, PhD, Kathryn E. Barnard, PhD, RN, Darcy B. Carr, MD, MS. Biological Research For Nursing, Vol. 8, No. 3, 210-222 (2007)

274   Maternal stress and behavior modulate relationships between neonatal stress, attention, and basal cortisol at 8 months in preterm infants.Mai Thanh Tu, Ruth E Grunau, Julie Petrie-Thomas, David W Haley, Joanne Weinberg, Michael F Whitfield Dev Psychobiol. Mar 2007 (Vol. 49, Issue 2, Pages 150-64)

275   Clinical efficacy of target low-intensity laser radiation on the adrenal projection region in patients with bronchial asthma. A V Nikitin, L A Titova   Ter Arkh 2006 (Vol. 78, Issue 3, Pages 39-40)

276   Intravenous laser irradiation of the blood at the health-resort stage of treatment for bronchial asthma patients. Iusupalieva MM, Shatrov AA, Grabil'tseva TA. Vopr Kurortol Fizioter Lech Fiz Kult. 1996 Sep-Oct;(5):9-12.

277   Low Level Laser Therapy restores the impaired relaxation induced by TNF-a of rat trachea smooth muscle via increases of cAMP. F. Aimbire, J.M. Bjordal, V.V. Iversen, R. Albertini, L. Frigo, M.T.T. Pacheco, H.C.Castro-Faria-Neto, M.C. Chavantes, R.M.Labat, R.A.B. Lopes-Martins. Photomedicine and Laser Surgery. August 1, 2007, 25(4): 315-367. doi:10.1089/pho.2006.9990

278   Temperature-Controlled 830-nm Low-Level Laser Therapy of Experimental Pressure Ulcers Raymond J. Lanzafame, Istvan Stadler, John Coleman, Belinda Haerum, Philip Oskoui, Megan Whittaker, Ren-Yu Zhang. Photomedicine and Laser Surgery. December 1, 2004, 22(6): 483-488.

279   The Effect of Low Level Laser Therapy (LLLT) on Stage III Decubitus Ulcers (Pressure Sores); a Prospective Randomised Single Blind, Multicentre Pilot Study. C. Lucas1, C.H.M. Coenen1 and R.J. De Haan. Lasers in Medical Science. Volume 15, Number 2 / March, 2000

280   Exposure of the volunteers' skin to visible polarized light is accompanied by increase in growth promoting activity of the circulation blood for human cells in vitro: Dependence of the effect on cell type. K.A.Samoilova, M.I.Blinova, N.M.Yudintzeva, N.V.Kalmykova, E.V.Kuz'minych *D.I.Sokolov, O.I.Zubanova, K.D.Obolenskaya

281   The use of low-intensity laser therapy under sanatorium-health resort conditions in proctology patient.Matsiak IuO, Varyvoda IM, Slobodian IeR, Mariuk IV. Lik Sprava. 1993 May-Jun;(5-6):150-2.

282   Verrucae Pedis: Evaluation of a method of eradication using class 3B (Low-Level) Laser Therapy – A double blind clinical study. Turner, W. (1997) Chiropody Research and Information notes, Omega Laser Systems. London.

283   A Preliminary Pilot Study in the use of Low Power Laser as a Noninvasive Technique for the Treatment of Plantar Warts and Verrucae. Burrows, V. Dyson, M. and Salah el Sayed (1990) Omega Universal Technologies. London

284   A Non Invasive Laser Management of Warts and Verrucae, A comparison Study Between Three Probes. Burrows, V. Dyson, M. Sayed, E. and Diamantopoulos, C. (1991) Omega Universal Technologies London

285   Effect of laser irradiation on latency of herpes simplex virus in a mouse model. D Perrin, J R Jolivald, H Triki, A Garbarg-Chenon, B Lamotte D'Incamps, B Lefevre, G Malka, L Miro, J C Nicolas Pathol Biol. Jan 1997 (Vol. 45, Issue 1, Pages 24-7)

286   Low-intensity laser therapy is an effective treatment for recurrent herpes simplex infection. Results from a randomized double-blind placebo-controlled study. A Schindl, R Neumann J Invest Dermatol Aug 1999 (Vol. 113, Issue 2, Pages 221-3)

287   Raydiance and FDA Sign Cooperative Research and Development Agreement 7/16/2007. PETALUMA, Calif., July 16 /PRNewswire/ -- Raydiance, Inc., developer of the world's first compact, cost-effective and fully software-controlled UltraShort Pulse (USP) laser system, today announced that it has entered into a two-year cooperative research and development agreement (CRADA) with the U.S. Food and Drug Administration (FDA).

288   Experience on treatment of children with otorhinolaryngological diseases by low intensity laser irradiation. A I Gogeliia, N M Chabashvili, M K Matiashvili, P A Kasradze. Georgian Med News Jan 2006 (Issue 130, Pages 84-6)

289   Low-intensity laser irradiation in otorhinolaryngology. M S Pluzhnikov, A I Lopotko. Vestn Otorinolaringol (Issue 2, Pages 5-14)

290   Poetry Corner, L DelRe, Mensa Journal 1987.

Made in the USA
Lexington, KY
13 January 2013